FIGHT FOR THE QUANTUM

ESSAYS ON SPIRITUALITY AND SCIENCE

JOHN LaCASSE

Mechanicsburg, PA USA

Published by Sunbury Press, Inc.
Mechanicsburg, PA USA

www.sunburypress.com

For information about special discounts for bulk purchases, please contact Sunbury Press Orders Dept. at (855) 338-8359 or orders@sunburypress.com.

To request one of our authors for speaking engagements or book signings, please contact Sunbury Press Publicity Dept. at publicity@sunburypress.com.

FIRST SUNBURY PRESS EDITION: February 2026

Set in Adobe Garamond | Interior design by Crystal Devine | Cover by Lawrence Knorr | Edited by Barbara Kindness.

Publisher's Cataloging-in-Publication Data
Names: LaCasse, John, author.
Title: Fight for the quantum : essays on spirituality and science / John LaCasse.
Description: First trade paperback edition. | Mechanicsburg, PA : Sunbury Press, 2026.
Summary: *Fight for the Quantum: Essays on Spirituality and Science* is a provocative and playful exploration of the philosophical battleground where science and spirituality collide. Rather than offering definitive answers, LaCasse invites readers into a lively polemic that dances through the "hard problem" of consciousness—the question of why physical states give rise to subjective experience.
Identifiers: ISBN : 979-8-88819-374-7 (softcover).
Subjects: SCIENCE / Physics / Quantum Theory | PHILOSOPHY / Metaphysics | PHILOSOPHY/ Movements/Phenomenology.

Designed in the USA
0 1 1 2 3 5 8 13 21 34 55

For the Love of Books!

Dedicated to the memory of my two boys

John LaCasse 1964–2012

Jeff LaCasse 1968–1989

And their remarkable brothers

Bob LaCasse

Erik LaCasse

"The two most important days in your life are
the day you are born,
and the day you find out why."

—MARK TWAIN

Contents

Author's Note

Richard Dawkins wrote *The God Delusion*, then David Berlinski answered with *The Devil's Delusion*. Deepak Chopra and Leonard Mlodinow got into a pissing match and decided to fight in the same book, *War of the Worldviews: Where Science and Spirituality Meet—And Do Not*. Michael Shermer writes his history tome in *The Moral Arc*. Sean Carroll tells us how it is in *The Big Picture*. Rupert Sheldrake debunks the constants of nature in *Science Set Free*. Yuval Noah Harari reminds us in *Sapiens* that if it were not for our (Sapiens) ability to contrive abstract myths, we would, in fact, be monkeys. Eben Alexander goes to heaven, gets amped-up by God, and writes many books. His Holiness, the Dalai Lama, reads all the books and comes up with three options. Phenomena are either (a) evident (b) hidden (c) extremely hidden.

So, who is the smart one in that pack? No wonder the Chinese government wants to control the Dalai Lama's 15th reincarnation. This guy is good!

In *Fight for the Quantum*, we discuss the interplay between science and spirituality as a polemic. Sometimes subtle, sometimes not so much. In all cases, the issues are obscure. That's why thinkers call this stuff the "Hard Problem."

In Josh Weisberg (2019),

> The hard problem of consciousness is the problem of explaining why any physical state is conscious rather than nonconscious . . . The usual methods of science involve an explanation of functional, dynamical, and structural properties. This suggests that

> an explanation of consciousness will have to go beyond the usual methods of science. Consciousness, therefore, presents a hard problem for science, or perhaps it marks the limits of what science can explain. (The hard problem).

There are no answers in this book, only observations of the many who claim to have answers. It's a book about the folly of the human condition. It's the kind of book a person can enjoy from any perspective.

Rupert Sheldrake put this in context after I sent him some early material: "John, I'm glad you're writing about these topics and your style is lively and engaging, but it definitely needs tidying up a bit." (Is he British or what?!)

He suggested six changes, and I made them immediately. That's the hard problem with this book. It wonders and wanders in everyone's context. It does its own quantum jumps among the Dalai Lama's three options. But, it's a hell-of-a-fun-read.

Foreword

In a book that has no answers, any reader will exit these pages with a sense of accomplishment. When Dr. LaCasse told me I was to write a foreword for this book, I was—and frankly, still am—confused. Regardless, as from the perspective of not the expert but rather the student, here goes.

This is a book that not within the first three hundred words declares its stance as omniscient, to which it releases us to submerge completely and without constraints. As the narrative drifts seemingly without continuity from stories of God to its latter, the imagination drifts with it. And so does our comfort. Stories of people who are smarter than us doing smarter things we can't understand will at times be discouraging. But it is with the regular comedic anecdotes and thorough explanation that Dr. LaCasse enables us to read with confidence. He covers foreign ideas with precision, all while developing an undertone that somehow remains to be open-ended.

The worth of this book is supplied by the randomness of the stories included. It relies less on compounding concepts but rather ideologies in short, digestible snippets. Meaning for the average reader, for every concept that seems dull, another thereafter will feel invigorating. Read this book, and you may find consolation in the cosmos or just a new-found interest. You will develop an understanding of quantum entanglement and will discover a new lens to view yourself and others. You will gain sympathy spawned from a common struggle and will look to the stars with, if not more confidence, more curiosity.

Shiah Sarkowsky

The Start

Apollo 8, the first manned mission to the moon, entered lunar orbit on Christmas Eve, Dec. 24, 1968. That evening, the astronauts—Commander Frank Borman, Command Module Pilot Jim Lovell, and Lunar Module Pilot William Anders—held a live broadcast from lunar orbit, in which they showed pictures of the Earth and moon as seen from their spacecraft. Said Lovell, "The vast loneliness is awe-inspiring, and it makes you realize just what you have back there on Earth." They ended the broadcast with the crew taking turns reading from the book of Genesis. (NASA) Borman: "And from the crew of Apollo 8, we close with good night, good luck, a Merry Christmas and God bless all of you—all of you on the good Earth."

It is interesting that while nesting in the hands of Newtonian hard science, the Apollo 8 crew ended the broadcast in soft science. Maybe toward the end, they turned, one-to-the-other, and said "We have a decision problem based on Turing's theory of computability. We must end this moment on a Riemann zeta conjecture. Do we have computable or incomputable sets?"

Or, maybe they didn't. Maybe the crew was so jarred by their witness that they deferred to a higher order. Maybe they turned one to the other and came with, "Oh my God, do you see that?!" as opposed to "*a giant leap for mankind*."

There are more than a few reasons why people defer to God. In new data on behavior research, deferrals could be considered conditioned reflex responses (Apr 2018: p.41). Like the actuality of Apollo 11 Astronaut Michael Collins when he says: "Jesus Christ, look at that horizon. Goddamn, that's' pretty! It's unreal." All of this, of course, before the crew opened the mic to the "good Earth" 238,900 miles away (Parry, 2009).

Like the blurt from Robert Ballard upon first seeing the Titanic. That capping off a fortuitous break from his search for the USS *Thresher*—a Navy *get there before the Russians* idea.

"I'll be God Damned!" he exclaimed. However, then, according to Ballard, he and his crew aboard the French research ship *Le Suroît* went stone silent (CBS News, 2018). It seems God managed to grab them by the nape of their necks, reminding them, in Ballard's words: "Nobody takes a shovel to Arlington, and nobody takes belt buckles off the USS *Arizona*." The crew was, at that moment, in the light of the Almighty.

Most everyone has enough intuition to understand; there is no upside to angering God. Luke, of "Matthew, Mark, Luke and John" fame gives us a rundown of God's benevolence:

> In Luke 12: 22-34 [Dependence on God] He said to [his] disciples, "Therefore I tell you, do not worry about your life . . . for life is more than food and the body more than clothing . . . notice the ravens: they do not sow or reap; they have neither storehouse nor barn, yet God feeds them . . . notice how the flowers grow . . . they do not toil or spin . . . do not be afraid any longer . . . your Father is pleased to give you the kingdom . . ."
>
> But, that blade cuts both ways: Deuteronomy 32:35, "It is mine to avenge; I will repay. In due time their foot will slip; their day of disaster is near, and their doom rushes upon them."

Distinguished Thinkers Are Idiosyncratic

Astronauts and explorers notwithstanding, great thinkers in all disciplines tend to be quirky. People deflect uneasiness with God, like cats scratching their ears to suspend stress. This wasn't lost on Einstein. According to K.C. Markides, in the Stanford Journal of Transpersonal Psychology, "Einstein's entire discussion provides a solid historical perspective on the emergence and development of Einstein's religious credo, that "science without religion is lame, religion without science is blind" (2014, pp.126–129).

There is more to this than the metaphysics of conversational response, however. Juxtaposed on both sides of the argument between skeptics and believers there comes the 11th-hour uneasiness about their thought processes that lead to tragedy. The brightest minds among the brightest operators seem to mire in the frailty of the human condition.

For example, Georg Cantor [Set Theory] resides among the world's extraordinary mathematicians. Cantor was born in Saint Petersburg, Russia. His mother was a Roman Catholic Russian; his father, a Protestant. This deformational conflict set the tone for his life. German fault finders berated him as he questioned assumptions of orthodox mathematics. His foundational crack might be Aristotle. Aristotle required specific confirmation. His assumptions led to the conclusion that only enumerations of finite sets were possible. From this, Cantor was never able to justify an infinite deity/god mathematically and finished his life in an insane asylum. Conversely, David Foster Wallace, the fellow who cracked the

Nazi "Enigma Code," was famously successful—yet, he put himself out of service with a Cyanide-laced apple.

Occasionally, wasted greatness has nothing to do with thought; we just underwrite stupid. The originator of Group Theory, Évariste Galois, managed to set his 20-year-old coition urges on Stéphanie-Félicie Poterin du Motel, a 17-year-old woman who was fully engaged with a senior fellow. So, in the opening stages of Galois's brightest mind, he manages to get crossways with her older suitor, they end up in a duel, and Galois takes a flintlock ball to his head and bleeds out under a few trees inside Paris. He died in Cochin Hospital the following day. There is speculation among historians that he was assassinated because of his political activities. He was arrested twice, thereby becoming a problem for the Paris police. Some believe the cops killed him, making it appear to be a duel over the honor of dear Stéphanie Poterin.

More recently, the urban phrase *I want to get away from it all* carried the day for a premium thinker widely respected in France as "the greatest mathematician of the 20th century." He was Alexander Grothendieck and his synapse of mathematics higher-order thinking dissolved into flowers. Praised by President François Hollande of France as "one of our greatest mathematicians" and "an out-of-the-ordinary personality in the philosophy of life," Grothendieck turned away from mathematics when he was on top of his game and retreated to the French Pyrenees as a hermit in seclusion.

The Plan for this Book

This study will drift between thinkers from actuality to abstention. We will review comments from the visible players: Richard Dawkins, Deepak Chopra, Stewart Hameroff, Sam Harris, Neil de Grasse Tyson, Rupert Sheldrake, Roger Penrose, and beyond.

The capstone will be examinations in the near-death experience of Neurosurgeon Eben Alexander. He is the man who dropped into a coma, went to Heaven for a week, drifted around with a butterfly, met his already-dead sister, and returned on the command of his back-on-Earth son (D.E. Alexander, 2012).

Then there is Libby Kelleher-Carr. The woman who, while on an assignment for the Sierra Club, showing three members of Congress the

Snake River Gorge, managed to flip upside-down and underwater long enough to spend the next several years elucidating whether that was a psychotic break or a conversation with God (Libby Kelleher-Carr, personal communication, January 11, 2019).

And, Don Martin, C-Suite Executive and lawyer, whose astounding vision in the arrow of time maps onto his life, changed his original context forever (Don Martin, personal communication, February 27, 2019).

As a preface endnote, on a special night in Seattle, my boys and I sat in the immediate proximity of Stephen Hawking. We were invited back to engage with Hawking on matters of complex cosmology. As we looked upon the man holding Newton's chair as Lucasian Professor of Mathematics at Cambridge University, we looked back at each other in silence. In the morphic field of my conscience, I uttered the following, "Oh my God, I can't believe this is happening!"

The Enigma of Sentient Beings

During a Russian purge in the 1930s, the Lykov family fled into the Sayan mountains of Siberia, Russia, 160 miles from civilization. Today, Agafia Lykova continues living at this secluded site. Agafia works in daylight. She tends livestock, fishes, cuts firewood, and listens to nature. She was born in this cabin and found her life rewarding.

Agafia is religious. Each morning she opens what her family escaped with—a 400-year-old book—and prays. She does not read the book; she holds the book. She completes this daily ritual in front of a small window as the sun rises. She knows that out that window, sentient beings are waiting for her to walk into the morning light. She knows that these beings are curious about her, sometimes afraid of her, and occasionally would attempt to kill her for food. She understands, and operates, in the balance of these relationships.

There are those among us who would challenge her prayers as unfounded; that her book is a useless artifact of an uninformed time. That the God she prays to is nothing but the stuff of thought; that she would make better use of the book by using it to start fires.

Moreover, there are those among us who believe she has established a resonance with the universe familiar to all existence, believing that daily rituals are in concert with all that exists, including doppelganger universes measured in billions— maybe trillions.

In an urban household, a person gets out of bed at the ring of an alarm. Simultaneously, that person's cat, dog, or pet companion of some kind, dances as they collectively, and in unison with the person, move through the house. They look at each other with repeated glances. They calculate where each is in their contrivance of superposition. They anticipate each other's moves and act accordingly. They exchange recognizable vocalizations. They are part of a nuclear family.

In the first case, occupied by Agafia Lykova, the environmental balance of her life causes her to reflect and react. Her dance is with the unique interpretations of her fellow sentient beings in the forest primeval. During the span of her day, she does not question the dimensions of her universe. She coexists with that universe. She lives without the knowledge that much of the light in the night sky originated 10 billion light-years earlier. When she lays out some pine nuts for a bird or squirrel, she is not calculating the homeostasis correlation between herself and the squirrel as the new frontier in Human-Animal Interactions in Zen. No, she is just there with a few pine nuts for the bird and the squirrel.

In the urban household example, the same relationship plays out in a similar ritual. The gets-up gets-ready person is dancing with a dog and cat as if working a soccer ball down the hall space from bedroom to kitchen. There are muted cries and wagging tails in penetrating anticipation. As the person begins to solve the food problem for the cat, there is no discussion between them regarding reflective self-inquiry. The dog is not practicing breath awareness. When the morning food hits the floor, the household person does not remind the animals of their afternoon mantra meditation.

Theosophical Tension in Science

At least this no entanglement/meditation of man and animal is the position of polemicists like Richard Dawkins, Michael Shermer, and Sam Harris, who are invested in material existence. It did not work for uncompromising players like Karl Jung either. His collective unconscious connections were only between members of the same species. So, these indissoluble science people see no conscience connection that resonates between pets and their owners. Each of their brains is only drawing from previous experience on how to react to the other's signals. And their previous experience is completely random. It's all random, with the sophistication of the brain making sense of the patterns. Neuroscientists know this through identifiable parts of the brain that excite against known stimuli from controlled experiments in relatable circumstances.

So, how about the Platonic theology people—the soft science side? They come from Marsilio Ficino's 18-book masterpiece, *Theologia Platonica de immortalitate animae.* He argues that there is a relationship to God and angelic beings on one side, and the qualities of bodies on the other (Dawson, 2014). The question being, are the bodies of Marsilio Ficino—our dogs, cats, birds, squirrels, and their companion owners? And, of course, what about God? Is there a crossover relationship between higher orders (God) and Man in [a] general sense? Then, the overarching question, does this thought experiment relate to bright people or foolish people? There is a great deal of tension between hard and soft science in that regard, and as we learned earlier, great thinkers are not universally dependable.

Space and Time

We also need to get some concepts straight. Let's begin with time. Where is Time in all that exists? For example, call up Andromeda Galaxy (NGC 224) and ask Cepheus or Cassiopeia what time it is. Do you want time from a philosophical perspective or time from a physics perspective? Do you want time in set theory, or do you want a block of space-time? How about the arrow of time? Oh. never mind, we will give time to you at two to the high fifteen. So, we will use 32,768 Hz. That's Quartz Crystal time.

What about space? How much space? Try calling Alpha Centauri to inquire if they have space. You know, like maybe a room for the night. Alpha Centauri is way-the-hell-out-there, so you will need to ask for space in the past, present, or future. For, you see, we move in space, and we progress in time. Ok, so you want space in block time. That would be four constructs—three in space and one in time. That way we will know where you want to be in space and time.

Space allows for independence. It can be hot here and cold over there. Time is relentless. I don't care if you are cold, you will move into hot on my terms, the speed of time.

Einstein and his ensemble of physicists needed time and space to try to understand the cosmos. We are not going to hector hard scientists and astrophysicists because they needed some understandable measuring to get through their process. For if it were not for these guys noodling through their thought experiments, we could not form the questions of the inquiring mind. Like what would happen between my cat and me if time stopped?

Do Cows Go to Heaven?

In 1893, John R. Daily, along with his partner James Walsh, started Daily's Premium Meats. They opened five retail stores in Missoula, Montana, specializing in beef, pork, veal, poultry, and buffalo. To support their business, they built a slaughterhouse along the Clark Fork River just west of town, on Mullan Road. The company, in various forms and locations, still exists today. The slaughterhouse, however, was closed in the 1980s.

When I was in grade school, in the 1950s, I used to walk down to the river and fish. That was a premium benefit of growing up in Western Montana—walking just about anywhere to do just about anything. An over-simplification maybe—or maybe not. I would fish downstream of the Daily slaughterhouse because the fish were enormous. They were well-fed from the slop that ran from the slaughterhouse kill floor into the river. When I began these fishing adventures, my interest was in the size of the fish and the catch. I was kind of youthfully immunized against the sound and the smell.

Years later, when I was living in Ogden, Utah, I took a temp job covering for a friend at SWIFT & Company. My job was to hot water hose-off the kill floor. There was a chain that ran along a ramp. The chain had hooks on it. There was a hydraulic ram gun that was positioned against the animal's head as each one came to the top of the ramp. If you saw *No Country for Old Men*, you will remember Javier Bardem using one of these things to kill people.

Then more hooks were secured, the animal gutted, and the carcass dredged through a brine tank to help detach the hide from the body. The smell was warm and thick, like being inside the body of a fresh kill.

When Javier Bardem approached people to kill, they were unsuspecting. The hydraulic strike was swift—no sound except the click of the trigger. The efficiency was unnerving. When the livestock gathered outside the slaughterhouse at SWIFT, there was a great deal of sound. The sound brought me back to the river bank in Montana.

The John R. Daily slaughterhouse ramp was outside the building leading to the top floor—kind of like a long, descending fire escape. The animals were compressed to enter the ramp one at a time. As they started up the ramp, they began to shudder in a high-pitched squealing cry. They lurched hard backwards, slamming their heads from side to side against the rails of the ramp. Workers moved them with electric prods that they jammed through the rails. The cattle would urinate and defecate out of control. As the animals listened to the others' cries, death was upon them.

The Hubris of Hard Science

René Descartes (1596–1650), the French philosopher famous for Dualism, considered among the intellectuals of the Dutch Golden Age—the "I think, therefore I am" guy—dissected his wife's dog by nailing the dog's paws to a board and cutting the dog apart without anesthesia to prove that animals are mere machines incapable of pain or thought. The sounds are simple mechanistic sounds. A test of his followers was to observe how hard they kicked their dogs. For, you see, dogs do not think. Therefore, they are not. Or, something like that.

Today, taking a position against Descartes Dualism, there is a belief that all self-organizing systems like atoms and cells increasing up the chain to the highest orders have the capacity for experience known as Panpsychism. And what follows is an Oxford University Press cited section that, by its existence, gives us the path to whether the high-pitched squealing cry of the cows on their way up the ramp and the bleeding yelp and shudder of Descartes' wife's dog represent a realistic variant leading to pain and fear.

In 1979, Thomas Nagel argued that if reductionism and dualism fail, and a nonreductionist form of strong emergence cannot be made intelligible, then panpsychism may be viable. It was not until David Chalmers's *The Conscious Mind* in 1996 that debates on panpsychism entered the philosophical mainstream. Since then, the topic has been growing rapidly, and some leading philosophers of mind, as well as some scientists, have argued for panpsychism.

Galen Strawson claimed that panpsychism is not an alternative to physicalism but is the only realistic variant. Panpsychism is the thesis that mental being is a fundamental and ubiquitous feature of the universe. (2016-12-29). Bruntrup, G., & Jaskolla, L. (Eds.).

Now for the obvious question: Did it take until 1979 for science seers to decide if the Descartes dog didn't like the idea of having his paws nailed to a board while being eviscerated by a French/Dutch dandy who liked to kick dogs? I suppose none of this should take away from Descartes's contribution to Dualism, and the worth of the pineal gland from which his greatness came forth in *Cogito, ergo sum.*

Descartes's obsessive posturing as tutor to Queen Christina of Sweden, the queen of keeping her environment extra cold, gave him pneumonia, which killed him. Riding on the back of chaos, creativity, and cosmic consciousness, I enjoy speculating on Descartes's relationship with his wife's dog, which I arranged to be [a] shape-shifted Wolf on the other side, just in case they meet again. Maybe, think of this in terms of the slaughterhouse.

Or, according to C. Koch, in his article "Ubiquitous Minds," in *Scientific American* (2014, p.26),

Why would God resurrect people but not dogs? This core Christian belief in human exceptionalism did not make any sense to me. Whatever consciousness and mind are and no matter how they relate to the brain and the rest of the body, I felt that the same principle must hold for people and dogs and, by extension, for other animals as well.

Major Players in the Fight for the Quantum

Frequently, people ask if Richard Dawkins has pets. In his public persona, he leaves little doubt that dogs and men go nowhere but into perpetual anesthesia when dead. On the other hand, maybe he would send his dogs ahead based on his relationship with Time Lady Romana. They were found to have a couple of dogs when paid a visit at home by *The Guardian*.

Dawkins and "Lalla" Ward (his wife 1992-2016) lived in a large house in north Oxford. Ward is an actress famous for her role as the Time Lady Romana in *Doctor Who* . . . Their two dogs, Cuba and Tycho, had the air of pets who run the joint. "Silly dogs," Dawkins said fondly, as they yelped and jumped at his knees (Elmhirst, Year 2015, 01.00 Edition).

Dawkins is the polemists' go-to guy for atheism, especially now that *God Is Not Great* author Christopher Hitchens is dead. According to Dawkins, all his writings should touch nerve endings, as did Carl Sagan's. The problem being, Dawkins tends to appear enamored with the gossamer side of nature, and therefore entangled spirits. As we just learned from *The Guardian*, in 1992 Richard Dawkins married Lalla Ward, who became the Time Lady. Indeed, she was Time Lady Romana in the BBC television series "Doctor Who."

> Romana and the Doctor (Who) arrived on the planet Zanak . . . Here they met the Captain, who used Zanak to materialize around and mine other planets . . . On her first visit to the planet Earth in 1978, Romana befriended Professor Amelia Rumford and Vivian

> Fay . . . The search for the fourth segment brought them back to Earth . . . They faced Millicent Ferril, a woman with the ability to control ferrous metal (Fandom, December 6, 2018).

Could it be that while engaging with Time Lady Romana, Dawkins was sending false flags to Sam Harris and Michael Shermer? Could Dawkins be a shadow theist? During his marriage to the Time Lady, was he walking with God in the multiverse? Was he making love to a time-traveling Seraphim? If there is a superhuman order of women time travelers, Dawkins can consult Deepak Chopra and Rupert Sheldrake on protecting himself with shrines and tree fairies. Then he can be coached into writing another book—*A Practical Guide to Marrying Fourth Dimension Spirits*—pet dogs notwithstanding.

Richard Dawkins is a person of global consequence and respected by friend and foe alike. However, like Sam Harris and Michael Shermer, he is a writer and methodical polarizer. In his polemics, he reminds me of U.S. Marine Corps General (Mad Dog) Mattis "Be polite, be professional, but have a plan to kill everybody you meet." Dawkins's academic title at Oxford was Professor of The Public Understanding of Science. So, he is a rhetoric and dialog guy for general consumption, not a scientist.

The same issue with Michael Shermer. He is a writer for *Scientific American*, edits magazines, and writes books. His Ph.D. is in the History of Science. If you put these three in a neighborhood Good Humor® truck marked "Hard Science," what you get is soft ice cream in Saran Wrap to appear hard.

I do not expect to see Dawkins singing Christmas carols during the holy holidays of Christ's birth. However, he could be in discussions with Neil deGrasse Tyson regarding direct evidence for the strange nature of dark matter. Tyson and Dawkins are on the same side of the God debate. Dawkins's style is among the best evidence of the Socratic Method I've witnessed. His out-of-the-box response to a challenger is Socratic: That's interesting, why do you say that?

He must know through experience that a string of questions will get the presenter in trouble in a couple of cycles. There are two exceptions—Deepak Chopra and Rupert Sheldrake. When Dawkins plays his Socratic card, both Chopra and Sheldrake return answers with steady speed through unending cycles. All sides in these conversations become incredulous through their enormous academic depth. The collective process ranks with the best debates on record. I feel confident Cicero did not stand in the Roman Forum and do any better.

A great deal is learned from conversation-style. The issue in these exchanges generally returns to the faith vs. evidence question. Sometimes referred to as the tough question. Dawkins rests on hard scientific evidence. The kind that is replicated in experiments. Evidence that can be measured in university science laboratories. And, certainly, nothing that falls outside the laws of scientific constants. To the Dawkins community, a constant is precisely what the term suggests. No change, no matter what!

Skeptic Sway over Spiritualism

Sam Harris was born on April 9, 1967, in Los Angeles. As part of his experience and education, he went to India and Nepal, where he studied meditation with Buddhist and Hindu religious teachers, including Dilgo Khyentse. In 1997, he returned to Stanford, completing a B.A. degree in philosophy in 2000. He received a Ph.D. degree in cognitive neuroscience in 2009 from the University of California, Los Angeles, using functional magnetic resonance imaging to conduct research into the neural basis of belief, disbelief, and uncertainty. Harris writes books and appears in media as a polemicist, rather than a scientist (cc Attribution-ShareAlike 3.0 Unported (CC BY-SA 3.0))

Sam Harris placed his education in the correct box for conversations between competing research science. However, he comes down quickly on the side of determinism. Interestingly, most peer reviews of Harris focus on his polemics. Like his contemporaries, he uses the mechanics of logic to form his arguments. Unlike Dawkins, however, who uses a Socratic approach, Harris leads with active voice. As a cognitive neuroscientist, he employs long engaging thought processes to develop the laddering of his positions. He is especially effective against any form of phenomenology. By using a sophisticated perceptual example, he can map-on to arguments in metaphysics, demonstrating where they dead-end in logic.

Harris plays these arguments as if he speaks for the people writ large calcifying them in his beliefs, through his own perfectionism, sometimes defensiveness, and intellectual muscle. He is most effective against academic theists who try to combat him with transcendental examples.

Harris fingered Deepak Chopra for being uninformed during a debate at Caltech. Chopra made the mistake of invoking the Quantum to explain universal consciousness. Harris publicly eviscerated Chopra for having the impudence to claim the quantum as a rationale for his position. Chopra fell four-square into the Harris logic trap. At the end of the conversation, there was an audience Q&A wherein Caltech's Leonard Mlodinow invited Chopra to take his class, so he would better understand physics and quantum field theory. This is more than a bit funny when one considers Harris in his self-imposed Determinism box.

Harris believes in [a] biologic predisposition of free will. He holds that there is a neuroscientific bias to a person's decision-making process thereby eliminating free will (Harris, 2014). So, any paradox available to quantum mechanics is unavailable to Harris.

Chopra is expressively closer to the quantum enigma than Harris, including Leonard Mlodinow, the theoretical physicist at Caltech, who entered from the crowd to inform everyone he was writing a book with Stephen Hawking implying by his own playing with Hawking in the Pantheon that Chopra was a fool. Chopra apologized for any misunderstanding and was pilloried again by Mlodinow.

Mlodinow and Chopra Monetize the Argument

Later, Chopra and Mlodinow (Have they no sense of creed?) decide to monetize that confrontation with a jointly written book *War of the Worldviews—Where Science and Spirituality Meet—and Do Not* (Chopra & Mlodinow 2012). The hardcover reviews arrived from luminaries, not least, His Holiness the Dalai Lama, who admonished readers and writers that, "We need a world view grounded in science that does not deny the richness of human nature and the validity of modes of knowing other than the scientific . . ."

Another reviewer was baffled by "well-respected persons" stacking accolades on the book with "apologies to the Dalai Lama, whom he assumed was seeking a higher perspective."

Unfortunately, the book was not the two contrarians mixing it up. It was a return to the dogma of each after having read the other's material.

It read like the editors at Three Rivers Press had laid down a marker telling Chopra and Mlodinow never to ask the question "Why am I writing?" The publisher wrote about the book, "Two great thinkers battle over the cosmos, evolution and life, the human brain, and God, probing the fundamental questions that define the human experience" (Chopra & Mlodinow, 2012, p. Cover 1).

I was looking for some Germanic mythology with a hammer-wielding god tossing around some thunder and lightning but, no . . . I ended up with High Tea at the Empress Hotel. Both these men are afraid of their position. It was like two politicians playing to their base. Chopra is working the heartland, and Mlodinow is playing the right—and left coasts. In one way, the book was a pleasant escape from the polemics between hard and soft science. Next time, I think Stephen King should ghost-write for both sides. Then their material would be fun.

Metaphysics and Art

For centuries the Earth has nurtured artists—people who believe the rationally mysterious. In the movie *At Eternity's Gate*, about the life of Vincent Van Gogh, American actor William Defoe recites a line from the artist, "There is something inside me, I don't know what it is. I see things nobody else sees." (*At Eternity's Gate*, 2018).

In *Fight for the Quantum*, we tend to pair hard and soft science at the exclusion of art. "A skill acquired by experience and the use of creative imagination" (*Art*, 2019). Now, let's change the subject in the sentence. In *Fight for the Quantum*, we tend to pair soft science and art, at the expense of hard science. A skill acquired by experience and the use of creative imagination. Doesn't that sound as valid as a skill acquired by factor analysis and algorithms? Does to me. And it resonates with 2.4 billion Christians, 1.8 billion people of Islam, and 1.15 billion followers of Hinduism. There are other typologies and breakaway groups that continue advancing these numbers in a linear progression.

Atheism as Spiritualism

Traditional atheists, like Richard Dawkins, accept the value of art in religion; in fact, they admire religious art. They do not, however, see a correlation to higher orders and religious art. Newer atheists, like Sam Harris, have a moderated view. I believe this moderated view is driven by the need to monetize their online empire, not because they are genuinely different from atheist tradition.

According to Sam Harris, "Human well-being is not a random phenomenon. The usefulness of religion—the fact that it gives life meaning, that it makes people feel good—is not an argument for the truth of any religious doctrine." (Brainy Quote, 2019). However, at walkingup.com, we can "Join Sam Harris—neuroscientist, philosopher, and bestselling author—on a course that will teach you to meditate, reason more effectively, and deepen your . . ." use of your phone because this exercise comes as a smartphone app. Is there some Sam Harris hypocrisy in this internet spiritual woo? I can foresee Harris introducing his next app: *To resolve everything, click here.* Nonetheless, he is well behind the curve because Deepak Chopra remains the king of *To cure everything, click here.*

Quantum Probability in Poaching Eggs

At the end of the nineteenth century, Max Planck slid a new thought under the door of Newtonian physics. Neils Bohr, Albert Einstein, and Louis de Broglie came along a little later adding more data. Finally, it was a graduate student, Hugh Everett, who provided the capstone of today's quantum field theory. Hugh Everett notwithstanding, what Planck suggested was that energy loss by charged particles came in quantum. Small subdivisions with multiple forms of energy. The field of energy for every particle was unstable (Universal Field over Particle eventually comes in Many Worlds Theory).

The way I see this is to imagine a pot of water on the stove. As you turn up the heat bubbles form on the bottom of the pot. Like when you are about to poach some eggs. Soon the bubbles begin to travel to the surface and pop. They disappear. It is impossible to predict where each bubble will break at the surface, but you know they will. There are only probabilities about the bubbles that include the whole surface of the water in the pot. You only know where the bubble is when you see it pop on the surface. There is a trilateral relationship between you, the surface, and the bubbles. And, as we will learn, the pot could be in superposition simultaneously bubbling in many worlds.

Albert Einstein, David Ben-Gurion, and Golda Meir were tight with the President of the New York City Garment Workers, with whom I have an oblique relationship. I have got to believe they kicked these poached eggs idea around off the quantum reservation as they aggregated the State of Israel.

In the end, Einstein, being a pragmatist, accepted the quantum by calling it spooky, thereby needing more study (Reuter & Saueressig, 2012).

Historically, not everyone was happy about this non-Newtonian happening; the least being Albert Einstein. He saw Planck's idea coming from uneven finite particles in light. Einstein's contemporaries viewed this as reckless. So, Einstein moved on to tweak Relativity—variable time and warped light. Much easier to think about as Einstein walked to work with Gödel, who kept telling Einstein that mathematics does not have all the answers.

Wherein Rests the Fight?

Today, the quantum vacuum fluctuation is appraised and arithmatized ad nauseam to suggest quantum field theory—that would be sans particles.

"Classical physics explains the world quite well; it's just the "details" it can't handle. Quantum physics handles the "details" perfectly; it's just the world it can't explain. You can see why Einstein was troubled." (Rosenblum & Kuttner, 2011, p. 7). Nobody has a solid answer beyond the probability of many interpretations, i.e., Einstein's needs more study. So, Chopra uses the uncertainty to suggest a lateral move into universal consciousness. Harris uses the uncertainty to suggest Chopra is on a jump-track to nowhere. Neither side will accept the other. It's like the NRA and Every Town for Gun Safety.

So, what are we fighting over? Perhaps the enigmatic reflection of our own bias. As Sapiens, we just need to know. Here's how the problem sets up: "A couple is in marriage counseling. The wife says, 'There is a problem in our marriage.' Her husband disagrees, saying, 'There's no problem in our marriage.' The marriage counselor knows who's right" (Rosenblum & Kuttner, 2011, p. 9).

Because we know this quantum story is metaphorically true, the spooky of this lucidity cannot be brushed aside. In our classical reality, debates on the credibility of quantum mechanics among deists and atheists will remain obscure until someone becomes believable when thinking outside the wire. That person might be Rupert Sheldrake

The "Outside the Wire" Boys

Depending on how you search for Rupert Sheldrake in Google, there will be well over a million hits. He is the master of educated calm in a very unsettled world. He is considered a quack by many classical scientists.

His TED talk on the constants of nature was banned (removed) from TED, then reapplied when his legions of followers hit the walls.

Rupert Sheldrake is a biologist and author of more than 85 scientific papers and eight books, and the co-author of six books. He was among the top 100 Global Thought Leaders for 2013, as ranked by the Duttweiler Institute, Zurich, Switzerland's leading think tank (Sheldrake, 2018).

The key atheist players are simply incredulous around Sheldrake. In 1993, Sheldrake was in a televised conference with Stephen Jay Gould, Oliver Sacks, Daniel Dennett, Freeman Dyson, and Stephen Toulmin. No light duty stuff there. His five companions reproved on Sheldrake's scientific worth personally and severally. Sheldrake was a visible token to the group. Toward the end, they let Sheldrake in for his summation of the collective conversation. He was polite and accommodating.

Then they wanted to hear of his current work—that they would have answers to his clutter. Sheldrake drew on his experience in Biology, then Ornithology, and called for the question: How do Pigeons find their way back to their coops? Game, Set, Match! It was over before it began. They were oblivious. They repeatedly tried to map-on blindfolds and other prosthetic devices to confuse the Pigeons. Sheldrake smiled, explained his research, and let them out of class early, to go home and ponder Sheldrake's research that put them communally under the table in less than ten minutes.

Sheldrake associated with Terence McKenna and Ralph Abraham and the triad became the silver chalice of original thought outside the wire. Outside the wire is military jargon for being beyond the safety of base camp. So, metaphorically, these men existed in enemy live fire. Before McKenna died April 2000, they collaborated on the book *Chaos, Creativity, and Consciousness*, and a series of lively trialogues, far-reaching discussions between them that took place between 1989 and 1998, in America and England. They all experimented with psychedelics from time-to-time.

Collectively, Google finds them today about seven million times in one-third of a second. These fellas would be the perfect Anti-Christ to Neill de Grasse Tyson, the people's physicist who walks the safe side of the boundary between consensus and controversial. Tyson is the not-in-my-backyard keeper of the classical physics keys. There is enough difference between the Tyson approach and the outside the wire boys, that if Sheldrake, McKenna, and Abraham were to walk into Tyson's Hayden Planetarium circa 1591 Anno Domini, they would be burned at the stake.

Aristotle and Acorns

This notwithstanding, there is evidence that Sheldrake believes Aristotle the Greek was right after all. Sprouting acorns do want to become trees. Considering this assumption doesn't fit neatly in the classical science worldview. We will explore a little of how we go back to Aristotle and try to remember the fundamental in-common construct of this book, that all the people I write about are smarter than me.

The philosophers of a previous period provided us with the platform of studying nature and the cosmos. However, the original Greek thinkers were inclined to take a pass on late-in-the-day deep thought. When the going got tough—usually just before Happy Hour—the truths became self-evident. The Greeks comingled politics with science. That way, tough questions could be argued during disambiguation. Positions became relevant only to the extent their aesthetic value fit the venue. It didn't hurt that all the questions and answers were mystical. Aristotle came before Newton so the Earth was still the center of the universe, wherein all must return. The scientific vernacular had the pentameter of incantations. The sun will rise, the cat will pounce, the moon will fall, and all the Gods like acorns. If there wasn't a physical explanation, woo would do.

The fallback position was biologic. Upon disagreement in the Basilica, there was a remedy. Take four ounces of blood from a wolf, two ounces each of a dragon's blood and brains, and one ounce of monkey fecal. Slip this mixture into your opponent's drink, and he will change his position.

The Great Chain of Being

Sheldrake takes from Aristotle's second assumption of universal connections. Aristotle was Plato's student, and the two of them chased each other around inside their heads. Thought experiments were Plato's card. He believed perfection existed in higher orders, and man could only approximate their pure excellence in form.

For example, God et al. held a perfect circle. Man could never draw a perfect circle. Aristotle got tired of Plato's thought loop and began to look to nature using syllogisms. He believed that circumstance brought a final cause. If a man swims, and a fish swims, then a swimming man must be a fish. Or something like that. Not to be confused with the anthropophagic phrase "Swims like a fish."

Most importantly, nature had a purpose. Aristotle developed a chain of being—The Great Chain of Being (Nee, 2005). God was at the top and plants at the bottom. The big news; they were connected. The mid-20th century German existentialist Martin Heidegger refocused that light once again in his writing, *Being and Time* (Agis 2014).

From this came Empiricism. Aristotle was the first *show me* person in philosophy. He thought about things he could see, as opposed to what he could only think about, e.g. Plato. Therefore, Aristotle began looking at nature, including animals, for signals to the meaning of existence. We can take this forward to where Renaissance science was Aristotelian. The interconnectedness of the great chain of being could be found in plants, and rocks, and the alchemy of pure gold from lead. The purer the human spirit, the better the chance of the chain connection between the created and God.

In the extension of Socrates, Plato, and Aristotle, we have the cradle of modern thought. They need not be taken in order because each context stands alone. Socrates was the arguer. Plato, the thinker. Aristotle, the mental mechanic. Except, watch out for Titus Lucretius Carus and De Rerum Natura!

Renaissance investigation expanded to Reductionism. As in: Let's disassemble this thing and find the explanation from the sum of its parts. The flaw here? Try this. Disassemble your computer. Lay all the parts on a table in front of you and, there it is. The internet!

So, the best we can do is establish a hierarchy upon which we can rest what we see and think. But, wait a minute! If there is no empirical evidence to what we just think about, there is no support. With that in mind, it didn't take long for proponents of nontraditional science to demand an elective seat at the table.

The primary New Age player in the public domain of holistic thinking is Deepak Chopra. Chopra believes a person may attain "perfect health," a condition "that is free from disease, that never feels pain," and "that cannot age or die." Seeing the human body as being undergirded by a "quantum mechanical body" composed not of matter but of energy and information, he believes that "human aging is fluid and changeable; it can speed up, slow down, stop for a time, and even reverse itself," as determined by one's state of mind. He claims that his practices can also treat chronic disease (cc Attribution-ShareAlike 3.0 Unported [CC BY-SA 3.0]).

Some physicists and astrophysicists' speakers treat Chopra like an airline cheap-seat hub-hopper. He is nowhere in their vernacular except delusional. Not far behind are Sheldrake, McKenna, and Abraham, but, Chopra is the absolute sovereign of the woo artists as he walks among believers in scientism like Sam Harris, Richard Dawkins, and Michael Shermer. However, in terms of raw market impact, Chopra is staggeringly effective. His organization has a reach that includes thirty-eight categories of mind-body experience through learning, events, and products (https://chopra.com/).

In Michael Shermer's book *The Moral Arc* and Richard Dawkins's book *The God Delusion*, Deepak Chopra is nowhere in either book's Index or Bibliography. Yet, Chopra is the elephant in the room for both Dawkins and Shermer. Chopra (1994) contends that the source of all creation is pure consciousness. He wants his followers to align themselves with the world's energy which he calls the "quantum soup" (p.8). It is this sniping of physics by using quantum for consciousness that grates on the contrarian team.

On YouTube, Michael Shermer opened a conversation with Deepak Chopra like this: "What-the-hell are you talking about?" (Shermer, 2012). Then Chopra, with a straight face, charmingly entered an explanation that mixed quantum physics with spirituality. Chopra knows what he knows, and no amount of traditional research will turn that around. But, here's the other side of that coin. Later we will explore the findings of A-listers Stewart Hameroff and Roger Penrose. Chopra might be closer than Shermer is willing to admit.

Michael Shermer and the Mata Hari Goats

Michael Shermer writes in *The Moral Arc* (Shermer, 2015) about Mata Hari goats. I read the book and then watched Shermer postulate about his position on science and reason. As a polemicist, Shermer is no Christopher Hitchens, but, for the sake of parity, as a pistol shot, I will never be found at the Camp Perry Nationals.

Michael Shermer is Richard Chamberlain in *The Thorn Birds* as he struggles with Maslow's Hierarchy (Zhao, 2014). He operates comingling moral positions to meet his complex horizon as public scholar and champion of argument and rhetoric. Of all the people on the speakers' circuit about the reason for atheism, he is the person most vulnerable.

Shermer calls for uncountable examples of Hobson's Choice as he makes his way through his confession of being a speciesist while exampling Nobel laureate Isaac Bashevis Singer.

"In his thoughts, Herman spoke a eulogy for the mouse who had shared a portion of her life with him and who, because of him, had left this Earth (Shermer, 2015).

> "What do they know-all these scholars, all these philosophers, all the leaders of the world—about such as you? They have convinced themselves that man, the worst transgressor of all the species, is the crown of creation. All other creatures were created merely to provide him with food, pelts, to be tormented, exterminated. In relation to them, all people are Nazis; for the animals, it is an eternal Treblinka" (Singer, 1980).

According to editorialists at The Berkshire Eagle; Pittsfield, Mass.,

Q: We know all too well about sexists, nationalists, racists. Now we have "confessions of a speciesist." What's this guy's bias?

A: He's *Scientific American* magazine columnist and book author Michael Shermer, who admits to being a "speciesist," or someone who presumes that nonhuman animals' interests are less important than those of our own species. "I find few foods more pleasurable than a lean cut of meat," Shermer confesses. "I relish the feel of leather. And I laughed out loud at the joke about the farmer who castrates his horses with two bricks: 'Does it hurt?' 'Not if you keep your thumbs out of the way.'"

Yet Shermer has sympathy with animal rights activists who equate "factory farms with concentration camps" and describes "the saddest slaughterhouse footage ever, showing a bull waiting in line to die." Hearing his mates in front of him being killed, he backs up into the rear wall of the metal chute and turns his head around, seeking an escape. A worker prods the bull along until the final death wall comes down behind him and down he goes in a heap. When an undercover USDA inspector inquired about the waste stench, one meat plant worker replied: "They're scared. They don't want to die." Shermer and I must have been fishing together at the John R. Daly slaughterhouse.

As Shermer concludes, "Mammals are sentient beings that want to live and are afraid to die. . . . Our genealogical connectedness provides a scientific foundation from which to expand the moral sphere to include not just all humans but all nonhuman sentient beings as well."

Within Shermer's work is the human code for context. I can see him in the Garden of Gethsemane. "Lord, forgive them for they know not what they do, and I like my flank steak marinated in soy sauce."

What the *Berkshire Eagle* misses with Shermer is his experience in the Galapagos Islands with Mata Hari goats. Here he runs four-square into his moral dilemma. In 2004, Shermer accompanied his friend and colleague Frank J. Sulloway on his expedition to retrace Darwin's footsteps. It seems the government of Ecuador wanted to keep the archipelago as original as possible. To that end, there were a bunch of interloping Mata Hari goats, and the goats were edging out the native Galapagos tortoises. There is speculation on how the goats got there but, no matter, the tortoises were there first and for thousands of years. So, the Ecuadorian government decided to kill the goats—all of the goats. While there, Shermer sees a young goat, gets close, picks it up, feels its sentient warmth, and slides into goat Nirvana (Shermer, 2015).

Just sayin', Deepak, your "friend" Shermer can be had!

The Ideology of Fractured Positions

In the secular camp, Dawkins (2006) holds that "our imaginations are forlornly under-equipped to cope with distances outside the narrow middle range of the ancestrally familiar . . . Our imaginations are not yet tooled-up to penetrate the neighborhood of the quantum" (p. 407).

Shermer (2015) feels any brand of spiritualism is xenophobic by nature . . . not seeking to embrace humanity outside their specific circle. "Religion, by definition, forms an identity of those like us" (p.153).

These comparisons amplify the disconnect between both sides of the hard question. The argument gets better. We have a well-known universal injunction "Love thy neighbor." In Chopra land, this is universal from the Torah meaning that it is the Law.

Love thy neighbor! Shermer (1994) lays down a different twist on *neighbor*. He feels neighbor is seen by religions as purely identical. If not identical, being a gentile in another county will do, but not with my kids and not in my community. What makes this even better is the reference for loving thy neighbor for deist and theist is from the Jews. And, from that camp, Moses assembled an army of twelve thousand troops to kill the Midianites bringing the booty and spoil to him. (Numbers 31:7-12). So, I guess the best we can suggest is invite your hog into the ring to see who gores who.

Physicians, Physicists, and Quantum Fluctuations

There appears to be a kind of oblique communication between physicians and physicists. Two high-profile physicians are willing to go deep with their interpretation of consciousness: Stewart Hameroff (anesthesiologist) and Deepak Chopra (internal medicine and endocrinology).

Hameroff with his hypothesis regarding microtubule, being supported in this option by physicist Roger Penrose, "known for his work in mathematical physics. In 1988 he won the Wolf Prize for physics, which he shared with Stephen Hawking for the Penrose–Hawking singularity theorems" (Wikipedia, 2019).

Hameroff and Chopra conversationally go after the 100 +/- billion brain switches that are unavailable in Ace Hardware. Stewart Hameroff sees high-end brain science (switching synapse on and off) as computation converting to a critical level of complexity, then leaving the observer in a kind of quantum smoke as the consciousness rabbit comes out of the hat.

Chopra buys that explanation as more support for his belief in universal consciousness. Both men agree that pure consciousness is included in space and time as a geometry of [a] block of space-time. Charge, spin and mass manifest in space-time for all that can be seen and unseen. Therefore, all of space and time are in probabilities. In our universal consciousness state, we pass through space and time as our observations trigger the actuality of our existence. All possible probabilities are present for our observations to actualize.

The rabbit out of the hat trick seems to come from microtubules inside the brain. They are structures inside neurons that operate in superposition to themselves. That superposition becomes the hat-trick. It's a now-you-see-it-now-you-don't quantum effect—like the rabbit sans the smoke. We are now down to the real deal in quantum mechanics. Things are separating with superposition influence at the primary—Planck scale geometry—structure of our conscious perceptions. And we don't know why.

"The term Planck scale refers to the magnitudes of space, time, energy, and other units, below which (or beyond which) the predictions of the Standard Model, quantum field theory and general relativity are no longer reconcilable, and quantum effects of gravity are expected to dominate." (Pikovski, Vanner, Aspelmeyer, Kim, & Brukner 2012)

Both Sides of the Hard Problem

Going forward, we must remember that the two sides of the hard problem around general relativity and quantum field theory disagree on outcomes. The traditional side holds that because the laws of physics are fixed, we live and die and that's the end of it. We are not connected with a universal consciousness because the atoms in our brains operate in our brains and connect nowhere else. That the atoms in our brains are incapable of activity outside our skull cap. And, anyone who believes there is a connection of any kind between universal consciousness and the atoms in our brain is practicing Quackery.

The other side of the hard problem believes that the quantum fields are the platform on which universal consciousness resides. That when research reaches the Planck scale level, there could be an atomic level happening that connects us to the body politic of the universe.

In popular media, traditional researchers hector anyone who believes there is more than alive and dead. So, man has, by today's standard, about 3 +/- billion heartbeats available, then he fades to black and stays there forever. No Heaven, no hell, no angels, no religion, no nothing. They accuse universal consciousness advocates of being uninformed non-scientists of questionable intelligence. Typically, the hectoring happens during a recorded event wherein the non-believer will leverage the logic of scientific method against the curiosity and faith of the other person. So, to make sense of what you read, you must be able to visualize what both sides see. Here is the secret sauce. They see nothing. The best either side has is interpretation through manipulation by equipment, or, the faith that the unknown leads to a dimension we are yet to understand.

The mechanics of atomic structure starts with the periodic table. In that table are the elements made of atoms. When we mix the elements, we get compounds that lead to stuff we understand like water and salt. The part of the periodic table that is in constant dispute is in the atoms Planck scale geometry. The part that is so small, we can only surmise what the equipment is indicating. And, of course, the equipment doesn't know either. To the best of my knowledge, God has not given anyone the code. The equation guys keep saying if it walks like a duck and quacks like a duck. And the harmonic guys say the only reason you think it's a duck is because of your narrow interpretation.

So, we are free to research as we like, read as we like, and practice our First Amendment right to free speech. Deepak Chopra, Rupert Sheldrake, and Terence McKenna like that. It gives Richard Dawkins, Sam Harris, and Sean Carroll high blood pressure. You may have spotted a bias in the last sentences. Terence McKenna is dead, however because he is on the soft science side, he still qualifies to be in their mix. That would be a universal consciousness. Just sayin' . . .

Spirituality Gets Its Nose under the Tent

The support for the Hameroff and Chopra conversation comes from the relationship between medicine and mathematics VIP's Hameroff and Penrose—both champions of respect—and the true nature of consciousness. Unlike most research relationships where civility and bipartisanship are excess weight to be thrown over the side, this triad accepts that consciousness remains unknown, but it does define our existence. Neither side is willing to fly the flag of the other. Penrose, for example is unwilling to recite any kind of spirituality. However, there is developing a framework edging toward universal consciousness from the Penrose-Hameroff theory of Orchestrated Objective Reduction "ORCH OR" (Hameroff & Marcer, 1998).

I will toss Rupert Sheldrake in here for the good of the order. Sheldrake is a been-there-done-that-got-the-T-shirt kind of guy when it comes to fields resonating simultaneously inside and outside the brain. He is in solid for a Morphic Resonance triumph in the face of the secular intelligentsia.

"In the 'Orch OR' theory, conscience events are terminations of quantum computations in brain microtubules reducing by Diosi-Penrose objective reduction ('OR') and having experiential qualities. In this view, consciousness is an intrinsic feature of the action of the universe." (Donadi, Bassi, Ferialdi & Curceanu, 2013)

Or put another way . . .

In perambulator terms advanced by Hameroff, the Orch OR theory describes networked microtubules as the rope in the Indian Rope Trick. These guys self-assemble (cytoskeleton) inside the brain neurons just because. Then they talk to each other through little windows called Gap Junctions.

But here is what makes all of this so interesting. They talk (communicate) during incoming synapses that cause the cell Axon to fire. The instructions come from the scaffolding-like protein network of microtubules something like this. Fire at will, however do not aim. Just point and shoot. In this random execution, the whole universe becomes their oyster. Put some here, and put a few over there, however, keep all your down range activity secret. Don't tell Brain Neuron Command what you are up to. Just do it. Higher orders will cover your tracks. If this were Headquarters Third United States Army, General G.S. Patton Jr. would go nuts.

There are several science arenas interpreting Quantum Mechanics. It seems every critical "seat" in the physics department of every important university has a thought on how to think about the quantum. A thought-provoking one comes from Richard Feynman's seat, being Sean Carroll at Caltech who supports the multiple universe ideas (many worlds theory) modifying particle theory with field theory. By reducing all of creation to a quantum field, all the spooky stuff gets more believable.

What Carroll advances supports Hameroff and Penrose in how the superposition in Planck scale geometry can exist in two worlds at the same time. In effect, superposition in space is everywhere and is not relegated to isolated incidents. For the original Schrödinger theory to work, all that exists is in, at least, a bilateral arrangement between worlds. The arrangement is probabilities, and man's participation in the quantum field with all he observes. However, as Carroll takes this further, for decisions to manifest, all probabilities of the final answer must be trilaterally in the universal quantum field at the same time, e.g. Schrödinger's alive cat, dead cat, and man's observation. Upon observation, when one doubles down on the prospect of an answer, does the field give up all the probabilities to a single outcome. And, to Sean Carroll credit, he prefers the cats to be asleep or awake, not dead or alive.

So, what this leads to is the unexplained landscapes of perception. In context, the average human brain enjoys a leveraged advantage to the average galaxy. That would be 100 million stars to the brain's 100 billion neurons. So, you see, the human brain has a computational leg up on the average galaxy. Silly maybe? Maybe not. There are some mass issues between stars and neurons. On the other hand, neurons have greater flexibility—just sayin'. :)

Consciousness in the universe uses the probabilities bridge between the quantum and classical worlds. Schrödinger thought the cat was dead and alive at the same time. I don't know of a story that gets more legs than that one. The magic in probabilities is occasion. As the synapses enter the brain neuron, they find the probable occasion to decide how to deal with this non-orchestrated command, e.g., Orchestrated objective reduction (OR). It's like a video clip from Dr. Who. The microtubules arrange in superposition. They place themselves to straddle classic and quantum physics at the same time. Then they wink at each other through the Gap Junction and trigger a microtubule automaton. An undefinable moment-to-moment structural geometry. This ponderable but elusive state allows consciousness to collate inside and outside the brain. At that flicker, we are connected.

How it works

You have a dog or a cat, maybe a bird. Your pet companion becomes sick. The vet says, "It's over. Time to put your dog down as the most charitable decision." As your dog dies in your arms, these are the questions that come through your tears. Does anyone on Earth know what my dog actually is? Does anyone know the relationship between my unexplainable dog and me the past 13 years? Why am I crying more over the death of my dog than the death of my father? Will my dog and I find each other again in the future? Same for your cat, and the same for your bird.

That's the odd thing about how we see science. It's perceived as secular. There is a definition of a dog in a medical codex. That definition is not close to defining my relationship with my dog. If my cat dies in my arms, under the same circumstance, does the last 3000 years of remarkably stable feline evolution describe missing my precocious little cat? No! Not even close!

These relationships are nested in the influence of how the universe exists. The universe exists in these moments of superposition consciousness. Every other event is a precursor to the moment microtubule superposition themselves in the cytoplasm as they support the goings-on around them. When conscious moments reach a numerical factor suitable to both classical and quantum physics, it is at that twinkling we reach the singularity of consciousness. We exist in this singularity of consciousness both in our brain and the interstellar, at the same time. So, when I hold my dying cat, our collective consciousness has this kind of zero-point oscillation exchange. Like an embrace in slow motion. As her consciousness begins to coalesce outside her brain, my consciousness is with hers as an overlapping seven circles grid with our combined seventh circle centered at our convening intersection. That's when the butterfly lands on your hand.

Panpsychism

Working in parallel, Sheldrake, Chopra, Penrose, and Hameroff operate in various states and strata of consciousness. Chopra utilizes the big picture, universal consciousness. Sheldrake sees causation as a hypothesis about objective observable regularities of nature (Sheldrake, 2009). So, species-to-species is a cleaner path than outside the inheritable factor. However, he does believe in a courtesy between all species created, including all of nature. Hameroff and Penrose see the potential for panpsychism in Orch OR theory. In separate conversations between these men, there comes a kind of Pythagorean harmony that underwrites both sides of the arguments revolving around the sub-atomic structure.

Sheldrake uses mechanism a great deal as examples of how that dead-ends in metaphysics. Penrose and Hameroff shy away from the stigma of metaphysics by merely advancing the notion that there is more to a mechanism than a computer can quantify. They cite Gödel's theorem as their example.

> According to Gödel's Incompleteness Theorems (1931), Incompleteness theorems are two theorems of mathematical logic that demonstrate the inherent limitations of every formal axiomatic system capable of modeling basic arithmetic. These results, published by Kurt Gödel in 1931, are important both in mathematical logic and in the philosophy of mathematics. The theorems are widely, but not universally, interpreted as showing that Hilbert's program to find a complete and consistent set of axioms for all mathematics is impossible" (para. 1).

Chopra cites in the blind with a question calling for a conclusion. His interview questions are interjections like a husband and wife who finish each other's sentences. The moment Chopra sees an opening he moves his comment into position. It's strategically brilliant how he can manage a conversation. Of course, he has years of practice. The Newtonian guys always have those micro-second delays to fact-check their response internally. Chopra advances straight at um. Then he delays waiting for a correction. Before his partner can contemporize an answer, Chopra is already at lock-and load for the next round. I think this is an evolution learning visited upon Chopra.

For decades researchers have been hammering on him for his belief in the worth of universal consciousness. Deepak Chopra has evolved as a wolf in sheep clothing. All nay-sayers, take notice, for Chopra might teleport himself to the Heaven's gate and set the conditions for your entry.

Sheldrake has a kind of lazy days of summer approach. Except, any inquisitor had better expect academic depth rivaling the Marianas Trench. Traditional researchers claim Sheldrake's experiments are anecdotally driven by his own bias. I would admonish the detractors to review how research works. Mixed methodologies with calls for an inquiry from the overlapping convention. Sheldrake, like all of us, hopes his findings will legitimize his hypotheses. What the other guys seem to forget is Sheldrake's enormously high profile mixed with an army of experimenters. There is no research upside for him trying to bullshit his legion of well-educated followers.

Sir Roger Penrose and Stewart Hameroff must meet the perceptions of hard science while working the trenches outside the wire. There seems to be a fundamental truth to research. There will be a multitude of interested parties trying to demolish the stages of inquiry. Rising above this are Penrose and Hameroff. On the ceiling of the Sistine Chapel, there is a representation of these two guys trying to decide if there is a spark.

Stewart Hameroff sat down with Deepak Chopra in July 2013 to discuss pure consciousness. The conversation focused on brain science and how the brain produces consciousness as a form of computation among neurons. Hameroff introduced Penrose into the conversation as his collaborator. More recently (December 2018) was Joe Rogan interviewing Sir Roger Penrose on Rogan's podcast, *Joe Rogan Universe.* Both Penrose and Hameroff corroborated each other's position on the validity of Gödel's higher plane mathematics. That there is something mathematics cannot factor.

Gödel was an associate of Einstein when both were at Princeton. Gödel was much lower profile than Einstein. However, Einstein afforded Gödel contemporary status with his understanding that Gödel was ahead of his math peers in orders of academic magnitude. There is a contemporary book *When Einstein Walked with Gödel—Excursions to Edge of Thought* (Holt, 2018), where we find a considerable discussion on how Gödel's influence was considered subversive in the academic community of Princeton. This was not lost on Gödel who, as we discussed earlier, became so paranoid, he began starving himself because he was afraid of being poisoned by rivals. In this effort, he managed to kill himself.

All this notwithstanding, the interviews brought to the public surface the extent to which Hameroff and Penrose feel they are on to the prospect of universal consciousness—to the absolute delight of Deepak Copra, and *I told you so*, Sheldrake.

Environment Decoherence in the Microtubules.

When Penrose wrote *The Emperor's New Mind*, his target was young people; a group the book might excite toward a career in science. Instead, the book gravitated to older people who had the time and energy to read 640 pages of scholarship. In this case, however, Penrose luckily caught the attention of a crucial senior reader, Stewart Hameroff, a research scientist and physician working at the University of Arizona where he spent most of his time as a professor in the Department of Anesthesiology and Psychology, and director for the Center for Consciousness Studies. Hameroff wrote Penrose a letter. Penrose, like all scientists of consequence, receives new idea letters with a great deal of frequency. This letter was different.

Hameroff could tell from Penrose's writing that he wasn't looking as thick as he might. He suggested Penrose look to the quantum effect of environment decoherence in the Microtubules.

"Tubulin, a flexible protein, assembles into a long chain to create microtubules. These 25-nanometer-wide tubes—thousands of times smaller than a red blood cell—are found in every cell in plants and animals" (Volk, nd, para. 1).

Penrose took notice and began to inquire about what Hameroff described. He realized microtubules in neurons were supporting the quantum collapse—the moment when scientists lost control of the mechanistic process.

As is always the case in research, Hameroff and Penrose's work/relationship was quickly denounced in the nonbeliever/skeptic community. This process mirrors the American political arena. Hameroff was producing fake news. Penrose was being duped by Hameroff who was unable to justify his claims. Their measurements were off the mark. What they suggested was impossible, and so on . . . This practice continues in scientific and consumer media. The only conflict I have yet to see reported is that Hameroff and Penrose are promoting a flat Earth measurement.

In Conversation

Stewart Hameroff & Deepak Chopra

In Chopra & Hameroff (2013r), Deepak Chopra sits with Stuart Hameroff and says, *Richard Dawkins and Michael Shermer are among my skeptics. These guys think you (Hameroff) are completely delusional. How would you answer them?*

Hameroff: *Well, they're wrong—they handle hard questions by ignoring them. They don't want to be bothered with details. For example, Daniel Dennett wrote Consciousness Explained where he rationalized away the problem. They look at the brain as a computing mechanism — no room in that for consciousness.*

Chopra: *In your opinion, is the universe conscious?*

Hameroff: *How about, its proto-conscious? It has pure consciousness built in without the snapshots of awareness. That only happens with this type of self-collapse or objective reduction.*

Chopra: *So, is the universe thinking?*

Hameroff: *Sure, why not, call it what you will.*

Chopra asks: *Could that be God?*

Hameroff: *You know I wouldn't have a problem with that. I think when you talk about God and spirituality, I think of three things. (1) There's interconnectedness among living beings which you can explain through quantum entanglement. (2) There's the ability to access and be influenced by cosmic wisdom if you will, divine guidance or following the way of the Dao. Penrose put these platonic values or proposed that these platonic values including mathematical truth, aesthetic, and ethical values are built into the universe.*

The Planck scale has patterns, and our conscious thoughts are influenced by the platonic values, so it's conceivable that we have influence by this cosmic wisdom if you will and the (3) third thing would be afterlife that we've talked about. The possibility that consciousness can exist after the body dies at least temporarily and perhaps be recycled in reincarnation (Chopra & Hameroff, 2013r).

Chopra exhales and says, *"There's cosmic wisdom, there's cosmic immortality, and there's cosmic truth, goodness, harmony, and evolution, and that's God!"*

Hameroff: *Well, we hope so.*

Sir Roger Penrose & Joe Rogan

Sir Roger Penrose and Joe Rogan have a conversation five years later:

Rogan In Penrose's (2018) podcast Rogan opens: *One of the things that I really wanted to talk to you about that I find quite interesting is consciousness and your belief that consciousness is not simply calculation, but that there's something more to it . . .*

Penrose: *Well, I mean we're all conscious, and so we may have theories about it. There's something else, and the something else goes beyond our current quantum mechanics, and it tells you what happens when the quantum state . . . doesn't follow the Schrödinger equation . . . and if there is something in the world which isn't something you could put on a computer, that's where it is . . . there's something non-computable, something beyond computation involved in our understandings of things.*

Stuart Hameroff pointed out to me these little things called microtubules . . . I had never heard of them at that time, but . . . I realized these microtubules are there, and they looked like just the kind of thing that could well be supporting . . . quantum mechanics up to a level where you could expect the quantum state to sort of collapse . . .

Rogan: *Are you convinced that microtubules are responsible for consciousness . . . ?*

And, like Hameroff, Penrose underwrites their 'Orch Or' theory.

Penrose: *"I think they're one of the best candidates . . . He (Hameroff) certainly thinks the microtubules are exceedingly important in consciousness, and I think he's right. That's the feeling I get."*

The Interconnectedness of Rupert Sheldrake

According to Deepak Chopra, "Rupert Sheldrake's contributions will be recognized one day on the same level as those of Newton and Darwin" (Sheldrake, 2009 Cover, 1). I think I would hang that in my office.

In the higher regions of favorite YouTube channels and personal appearances is Rupert Sheldrake. He began in science in an ordinary way, became an Atheist, changed that when he realized that to keep doing traditional biology, he would have to keep killing small animals, turned to Hinduism as he studied and worked in India, and returned to the UK. Eventually, he found himself listed among the top 100 Global Thought Leaders ranked by the Duttweiler Institute, Zurich, Switzerland's leading think tank.

Earlier in this text, Sheldrake appears in the dance with Atheists and researchers of all stripes who follow a mechanism path. Sheldrake seems fearless in the face of that opposition. He practically floats above his detractors. He deals in the no-see-ums of Biology, and Ornithology. He spins a convincing story about why the constants of nature are not constant. He holds that the constants are habitual. When speaking, he frequently uses a phrase borrowed from his friend Terence McKenna who observed regarding the first quantum of the universe, "Modern science is based on the principle: 'Give us one free miracle, and we'll explain the rest.' The one free miracle is the appearance of all the mass and energy in the universe and all the laws that govern it in a single instant from nothing" (Rupert Sheldrake 2018 p. xiii).

People who watch Brian Green on NOVA or read Sean Carroll's work on the state of the Universe, get to tango with the maybe yes and maybe no of the #1 quantum. That was the term used for the "primeval atom," by Father George Lemaitre SJ., Catholic Priest and Cosmologists. His hypothesis disparagingly referred to as the "Big Bang" by Fred Hoyle because Hoyle didn't believe it. The name stuck.

Sheldrake's position is "The sudden appearance of all the laws of nature is as untestable as Platonic metaphysics or theology" (Sheldrake, 2018 p. xiii). Sheldrake reminds us that Laws are of man. Earlier in this text, we skirt that idea when we briefly discuss the nature of time and space. The terms we use are constructs of man. Anything Un-Pythagorean is up for grabs. But we must keep in mind, the universe uses time, even though planets do not know what time it is.

Sheldrake tells a great story about a conversation with the official UK keeper of the speed of light. Over a period of years, Sheldrake noticed that the speed of light varied enough to make notation changes in the official record. When he approached the administrator in charge of such things; challenging him with the difference in the speed of light (through the same measuring medium for those among of us wanting to make the 'medium' case), the fellow exclaimed that Sheldrake had come upon a big embarrassment in the cosmology of light speed. When Sheldrake pressed the UK keeper of the speed of light for 'why the change in speed,' the fellow admitted that there are averages taken, that the speed was fixed by committee, and the process was known as, "Intellectual phase-locking" (Sheldrake & Shermer, 2016).

I guess somebody has got to make the call. What are we going to establish as the speed of light today? When Sheldrake brought this to TED along with nine other examples of the dubious quality of the constants of nature, his talk was removed from the TED index. His TED talk was banned. So, it seems to me the gravitational pull of the TED committee's determinism, or the UK's light speed keeper was more important than the (actual) speed of light. Always be mindful of a person wearing a plastic-breast-plate that reads "OFFICIAL." He may be responsible for establishing the second law of thermodynamics. The one about entropy, and, at the bar, you may never be able to make sense out of an *Arnold Palmer* again.

The Nature of Consciousness with Subjective Phenomenal Experience

People speculate on Schrödinger's equitation about being alive and dead at the same time. Sean Carroll prefers asleep and awake, and I agree. Schrödinger's daughter speculated that he didn't like cats. However, all the speculations we read work better if we have personal experience in space-time geometry to drive it home. We know chasing space and time together leaves vague answers. As we chase one, the other gets away. We know that Stewart Hameroff and Roger Penrose suggest there is something else, something hard science has been unable to find. We know that non-traditional soft science cuts us some slack. The best definition comes from artist Robert Rauschenberg: There are so many interesting ways to be other than right. So, at what combination of space and time do we experience both?

The most original examples come in emotion around death, or when someone has a Near Death Experience (NDE). Both are dramatic. In the case of multiple deaths through natural or manufactured tragedy, such as terrorism, the phenomena are expanded. If we are to believe the inquiry of Roger Penrose, Stewart Hameroff, Rupert Sheldrake, and Deepak Chopra, there is something more to death than finality. Sheldrake tells us that there is a species bias in Morphic Resonance. The morphic fields exchange easier between similar natural organizations of the structure.

Aristotle believed the entire system was top-down connected. From God all the way to bugs and plants. Sheldrake sees a positive field bias between the same similar species. Sean Carroll at Caltech says that there is one quantum field and it is everywhere, servicing everything. So, where is the moment in the quantum field of space-time geometry where personal experience gets the best traction for personal understanding?

Unforgettable Overlapping Consciousness

On February 2, 2019, there was a celebration of life for a woman who died too young. When alive in her body politic she lived with her husband at Diamond Point, Jefferson County, Washington State. The celebration was large with people filling the local Air Museum at Jefferson County Airport. Diamond Point is a small aviation community nested between larger population centers on the San Juan Strait, Kitsap Peninsula. The people who live there enjoy upper-middle-class success, including the freedom to fly their own equipment.

On this day, I accompanied one of the deceased's longtime friends to the celebration. My partner, being a member of this group during their early days of life and careers in Southeast Alaska. Back when fishermen and loggers were just that. When teachers and administrators were encouraged to go to Alaska and establish access to resources enjoyed through federal and state agencies, these people were part of that migration.

As we arrived, I noticed a cluster of people in full motorcycle leathers unfurling American flags, obviously to be held at parade-rest as more people arrived. I inquired and found they were the American Legion Riders. What am I witnessing here? Who are the American Legon Riders in relation to a dead woman whose group of friends were early adopters, and risk takers at the Last Frontier, in The King Salmon Capital of the World?

There were men inside wearing black wingtips with tassels circa 1985. There were $1,500.00 Hickey Freeman suits that still fit. There were baseball caps and insignia jackets on the flyers. There was food stacked upon food surrounded by cookies, flanked by gift baskets for all in the room. There was a continuous loop of family pictures moving through a projector toward a large screen. There were scores of sliced lemons drifting in glass water columns with silver spigots. Everyone was talking.

The afternoon opened with the food and closed with stories. The dead woman's husband gave a eulogy impressive enough that I am confident, after a few minutes, every person in the room was wondering if he might be available for their own funeral. Sitting among family were the woman's two daughters. It came their time to speak. In each case, they were unable to maintain a solid composure. They both resorted to reading messages they wrote earlier to their mother. As they, in turn, began to read I could feel, and see, a different spatial climate in the room. People were sitting on the edge. They were cocked to one side or the other. Some were leaning in. Some were twisted holding their heads with one ear to the front. There was no ease anywhere in the room. The room's atmosphere was being subjected to a kind of angular momentum. Every haberdashery marker was put silent.

There was about to be a singularity in the turmoil of each girl. If their mother's consciousness was at hand, it would be in this moment of space-time geometry. The girls detached from the crowd; they openly cried as they read to their mother. It was no longer the formatted eulogy. Their father was fighting tears. It was classic overlapping consciousness—their combined presence centered at their convening intersection. It was a Planck scale superposition in a nuclear family. Their quantum enigma gifted the room with its scientific and cerebral cachet. What we witnessed was an example of the nature of consciousness—a subjective phenomenal experience of internal and external worlds.

Manufacturing Strength

I grew up in the west—Montana, where winter temperatures are so cold, cars are kept running when they are parked downtown. It's where grocery stores have counterchecks at the check stands. It's where the 4th of July parade goes up and down the street, so people can see both sides of the floats. It's where the local police take you home to your parents instead of to jail. It's where you arrange for a minister to officiate at your wedding, and he shows up in full Native American regalia. You look at your father-in-law, and he says, "I didn't know Fred was an Indian."

"Totally cool," you say, and the night goes on till dawn.

For my growing up years in Western Montana, I hunted every fall with my dad—sometimes with my mom and dad learning how to track a deer, catch a fish, skin a bear, and lead a duck. When I left Montana, I drove all over my hometown, stopping to speak with some of my favorite places. The old irrigation ditch that ran through the south side. The beginning of the path to Hell Gate Canyon.

I walked up Mt. Sentinel to where I buried Buttons, my dog. I wanted some time with him, some alone time, so I could promise him I would remember him forever, and that I would be back someday, and that when I came back, he would be proud of me as I was of him. When I got to his spot, the grave was open. Most likely coyotes, maybe a wolf. Buttons was gone. I sat in the sloping grass and cried. I cried in pain, and I shivered in fear. Buttons was gone, and, as it turned out, so was my youth.

The State of Grace

What a firestorm of gut-wrenching pain this is going to be.

I was on my couch, a nine-foot Lucca light-brown saddle leather Henredon. It was a $15,000 idea I had years before. Silly idea maybe considering the depreciated value of furniture, however, for thinking, napping, occasional sex, telephone calls, and light reading, it was worth every nickel. John (my son) and I were having a phone conversation about exercise. He was a competitive sailboat racer (Santana 30) and felt that keeping in shape was an ingredient for success. There is a great deal of exertion involved with sail trim while holding off the mark and doubling down on speed for long periods of time. He was good at this, and it was clear he was one to be watched among West Coast competitors coming through the yacht club circuits.

Even though he was working hard to capture as much forearm and shoulder power he could develop, he was having difficulty with his right side. A friend noticed, as they were both working the cockpit during a race that John was having trouble working the primary winch on the starboard side. This was his power side and was a logical position for being quick and effective. And, yet, for all the practice, he could feel there was something up with his right arm.

Most people do not have a good reference point for limited weakness in their limbs. We come and go from ailments that are passed off as stress, or fibromyalgia, or just plain nothing except waiting for it to be better the following day. I also don't believe people tend to accelerate routine body pain to crisis proportions as a matter of routine. However, in John's case, there was another trigger that happened while he was sitting on the cockpit coaming after a race. He reached down to unscrew the crown on his Rolex so that he could adjust the date, and he was not able to manage his fingers to get that done. That was the trigger.

I didn't know any of this as it was happening. There was no reason for him to alert me. Enough time passed so John could consult with a physician team at Evergreen Hospital in Bellevue, Washington. From what I understand there was a great deal of testing, some remarkable, and some nonremarkable, as these things go. Finally, John and his wife Wen went in for a conference with the Evergreen medical team. The diagnosis was tentative because the physicians at Evergreen Hospital were not experts in this medical circumstance. However, they suspected that John might have Amyotrophic Lateral Sclerosis (ALS), commonly referred to as Lou Gehrig's disease. The recommendation was for John to transfer his case to Virginia Mason Hospital, in Seattle where there was a team of physicians studied in progressive neurodegenerative disease, specifically, ALS.

About a week after John got the diagnosis from Evergreen Hospital, and he made his way to Virginia Mason for a second consult, he gave me a call. He told me he had Lou Gehrig's disease and was wondering if I knew anyone who had it that might add some light on what to expect. His voice tone was low but controlled, and he seemed directed toward getting additional information from any available third parties.

What do you do when, after a one-sentence exchange, you know a person is going to die? When that person is your son. When you've been through this before with his younger brother. When you already know what a firestorm of gut-wrenching pain this is going to be. When children die first, it's never right, and I'm unaware of a circumstance where you can understand why. A dire death diagnosis, with pain and suffering to follow, is something that makes the road ahead look unbearable, so hopeless, that a quick self-imposed ending appears perversely easier.

As John and I spoke on the phone, there came over me nausea. I began swallowing for air and tightening my stomach muscles and in my mind, seeing him as my little boy. Oh, my God, my dearest little boy, my wonderful little boy. I remembered how he would wave goodbye to me, as he piloted his Santana away from the dock at Shilshole Bay and north toward Port Townsend. I would envision him walking toward my boat then, walking away. I once told him that sailing alone was like being a lone rider on a horse. The horse and the boat have a magic relationship with nature.

I was quickly losing my ability to sound normal, and I didn't see well through the welling tears in my eyes, so I told him I would get back to him shortly with any new information I could round up. I told him I loved him and lowered the phone. I sat up erect in my seat, cupped my hands into a fist, held it against my lips, and began to pull my arms in close to my body. I was forcing my arms in as tight as I could manage. I began to breathe in and out of my fist. I looked out the window to the sky. I stayed motionless. Stoic. Breathing.

I had been here before. In this state of grace, a place where our body's circuits begin to shut with staggering speed. The body's race to survive. The murky world of quick-time slow motion. The place where options end, and there is no choice but maintain course and speed.

In February 1989 (21 years earlier) I was asleep on my 54 Maple Leaf motor sailor at the Port of Seattle. Early in the morning the phone rang announcing what all parents consistently described as "their worst nightmare." My son Jeff was at Harborview Medical Center—he had been in a car accident.

My dark-of-the-morning trip to the hospital was filled with a general curiosity about his length of stay, the costs, whether it was his fault, and if others were involved. I could begin to see the lights of the hospital floors ladder their way higher as I drove up Boren Avenue. I parked the truck and found my way to the Emergency Room entrance. I walked quickly to the ER while announcing my name.

The eyes of the nursing staff began to dart all around—*if you stay right here for a moment, I will be right back*. She hurried into the hall and, in a few moments, came back with three green-gowned doctors. Their masks were each hanging off one ear; one of them was still gloved. The other was holding a piece of equipment. We all sat down. They jockeyed through the conversation. However, the message was they wanted me to know the situation was severe, that I must be prepared for a shock when we enter the critical care unit where he was being "positioned," and that they were prepared to support me at any moment.

As I am rifling this memory through my mind, I am taken with the contrast between the doctors in the ER for Jeff, and the matter-of-fact conversation I am having with John. I guess in Jeff's case, serious as it was, death wasn't on the table. In John's conversation, the fait accompli is death, right out of the box. The lead-in is desperation. So, I am masking how I feel, like an airline pilot on a direct approach into the side of a mountain. "Seattle Center, this is 311Quebec Heavy. It doesn't appear we are going to clear the north ridge."

The doctors and I walked toward the Emergency Room where Jeff was in isolation. As we entered, I could see him at the end of the room. He was tilted up, then his head would disappear, and all I could see were his feet. There was nursing staff moving seamlessly through the hiss and gasp of pump pressure that ran the equipment. I saw several beeping monitors with wires, some with tubes. One was alarming as it looked like a tracheotomy, and it was. He was secured to a metal table that was moving in kind of slow circular motion in a 4-axis routine—head up then down while simultaneously rocking from side to side. There was a stainless-steel halo ring around his head with long steel screws positioned into his skull. He was asleep.

There was no blood, no bandages. He was draped with a sheet, but a good deal of his body was exposed. His solid muscle tone stood in stark contrast to his restrained body. It seemed an odd dichotomy on a 20-year body.

One of the doctors said that they were unable to get an X-ray yet. However, they suspected he had a C-6 fracture at the base of his neck. One of the doctors turned to me. He said the special table was to "maintain hemodynamic stability and stabilize blood pressure values, as well as steady his spine." I imagine my eyes were trying to engage my brain as I watched him speak. Only later did a nurse tell me what he said.

As I stood next to Jeff, I had long since lost my ability to breathe involuntarily and was now moving air in through my nose, and pressuring it out through my pursed lips, desperately trying not to descend into hypoxia and drop to the floor. I was clammy and cold. I was shaking, and utterly terrified.

In both these cases, my sons died — Jeff in thirty-three days, John in about two years.

My last conversation with Jeff was at a moment when he opened his eyes as I was standing over him. He winked—his eyes hard-focused on mine, then he fell back to wherever he was. I left his side and walked to a window in the corridor. I could see far to the west, all the Olympic Mountains in high resolution. A nurse approached me. "You must come now," she said. Jeff was on his way. The equipment was flat-lining, then silent. I ran my hand through his hair, leaned through my tears, and kissed his forehead. I took some tape-scissors and cut away a lock of hair, stuffed it in my pocket, watched his body get covered and wheeled away—went to my boat, sat on the deck, and drank beer until I fell over.

My last conversation with John was by text message. I was at Staples making a Bluetooth exchange for my computer. John was unable to talk, but could text by using an electronic relay attached to his forehead.

Me: You Ok?
John: Yes
Me: ;-)
Me: Reinstalling Dragon and Service pack 1.
John: Which one?
Me: Calisto for Savi Go
John: Wish I could be there.
Me: Not the same without you.
John: Too bad, my Savi Go worked great.

The exchange stopped—Shortly, John began to choke and died of hypoxia.

Later that week I opened my phone and typed; I will love you forever.

By some measure, we are connected. Experience can walk on both sides of an imagination's line. I spent a great deal of time speculating on where my boys went. I have another two boys. They speculate as well. As a family, we use icons from those who died. Erik wears John's watch. Bob uses Jeff's guitar. I walk around with hair in my pocket. Among my personal friends, each of them has skin in this game. The boy's mother, she is in a class of her own—a very exclusive club.

It's a natural transition to cosmology when the search is on for life after death. As the socially constructed pressure of the 21st Century builds, there is accelerating interest in the other side. The opening of this book trots through examples and comparisons between various players operating in the authenticity of creation, the worth of spiritualism, and the impact of atheism. The examples are light, sometimes funny, and never without notice. All the speculations rest on space-time geometry, the playing field for the hunt.

Media and the Fourth Dimension

Among the chief facilitators of how to view our existence is Joe Rogan. He handles his YouTube channel with 4.5 million subscribers wherein he has conversations with the purveyors of existence in the seen and unseen.

At some point, Rogan, using well-managed Socratic exchanges, gets everyone to take a position somewhere in space-time geometry. He calls for the question, "*That's interesting. Have you heard of . . .*" and he leads the conversation toward understanding by the common man. Rogan is a gift to astrophysics, spiritualism, and cosmology. He is comfortable with both sides of the aisle. Rogan gives his "*What's up man—great to see you!*" opening to atheist Sam Harris while showering his excitement on spiritualist/scientist Rupert Sheldrake because of his trialogues with McKenna and Abraham. In each case, Rogan becomes the champion of the curious and information guide for the want-to-be informed.

The Size of the Field

In the biggest case, the universe is infinite. There was no beginning, and there is no end. The universe is because it is. That, of course, calls for the question, where does God fit in infinite? Or, who needs God, when the thing in question always was? No first cause. In this mix, we find German existentialist Martin Heidegger, flawed by his association with the Nazi Party, but still the top-of-the-heap-guy in existentialism. "*At the beginning of infinite, nothing met nothing where there was nothing.*" That, my dear readers, is Heidegger's definition of unoccupied.

In their intellectual tradition, astrophysicists use instruments to measure spectrum light from far, far away. They show us pictures of matter that, based on their place in the spectrum, came into existence billions and trillions of light years ago. In this picture, there are about 200 trillion galaxies. Or, about this many 200000000000000.

Each one of the 200000000000000 galaxies is +/- about 100,000 light-years across and 3000 light years thick. Keep in mind, the speed of light is about 186,000 miles per second. Or, (1= 5.875e+12). That is 7.5 times around the Earth in one second. You cannot say *1-2-3-4-5-7.50* fast enough to match the speed of light around the Earth seven and one-half times in one second. In contrast, the next time you board a commercial jet, you will cross the United States in 5 hours. It is reasonable to assume these numbers are hard to get our arms around. Least we say, there is a lot to work with, God or no God.

The Astrophysics Intelligencia of no God

Neil deGrasse Tyson, *I don't care what you believe about God, just don't believe it in my school*, lands solidly on the mechanistic side. Because he is a high—maybe the highest-profile—astrophysicist, there is a building trend away from spiritualism toward atheism. Tyson spins away from spiritualism toward Carbon and Hydrogen. Got to be a lonely place indeed, however, popular in a secular society. Tyson remains ambivalent about God to keep himself on a kind of show-me trajectory. He wants people who have a near death experience to ask questions and take notes—maybe pictures. Tyson is hegemonic. His *I am correct* position cuts a lot of people out of the pack. He may play to his base to keep his media profile active. He is enormously intelligent and pragmatic, however, as he examines the last 12000 years of discovery, he might be leaving himself open to criticism for not letting in anyone with a diverging experience as opposed to opinion. As we remember, for centuries the Earth has nurtured artists — people who believe the rationally mysterious.

To be fair, it's easy to understand how Tyson could get sideways with people's perception of all that exists. From his perspective, someone needs to snap-a-chock-line through the clutter. For example. If we visit Cosmology in Creative Commons Wikipedia, we find:

Hindu Cosmology

One cycle of existence is around 311 trillion years and the life of one universe around 8 billion years. This Universal cycle is preceded by an infinite number of universes and to be followed by another infinite number of universes. Includes an infinite number of universes at one given time.

Jain Cosmology

Considers the Loka, or universe, as an uncreated entity, existing since infinity, the shape of the universe as similar to a man standing with legs apart and arm resting on his waist. This Universe, according to Jainism, is broad at the top, narrow at the middle and once again becomes broad at the bottom.

Pythagorean Universe

At the center of the Universe is a central fire, around which the Earth, Sun, Moon, and planets revolve uniformly. The Sun revolves around the central fire once a year; the stars are immobile. The Earth in its motion maintains the same hidden face towards the central fire. Hence it is never seen. First known non-geocentric model of the Universe.

Milne Universe of Kinematic Relativity

Rejects general relativity and the expanding space paradigm. Gravity not included as initial assumption. Obeys cosmological principle and special relativity; consists of a finite spherical cloud of particles (or galaxies) that expands within an infinite and otherwise empty flat space. It has a center and a cosmic edge (surface of the particle cloud) that expands at light speed.

Reaching Across the Pawn Ticket

Leaving all of this in the queue for a moment, let's examine an experience, which operates well in the entanglement of space-time geometry.

When Jeff died, we delayed the funeral by a week so friends and family could arrive in Seattle. I was emotionally occupied by needing an answer to his being ok. The being ok part does not go away at death. Things that go wrong in this spanner period between alive and dead, not least, cooperation from disinterested third parties. Jeff had hocked his guitar. His girlfriend had the pawn ticket. I wanted his guitar for the funeral. When I went to the pawn shop, the owner told me, No pawn ticket owner, no guitar. I explained my problem, and money didn't help. He didn't care. I drove away. Later in the hour, I called the Everett Police Department. I got a detective. I told the detective that I was going back to the pawn shop, I would ask for the guitar, and if I got any more bullshit from the owner, I would break his fucking neck and take the guitar anyway.

The detective said STOP! We will handle this. He said, at a time to be decided, I was to meet the police in the parking lot of the pawn shop. Give them the pawn ticket and wait in my truck. I called a friend, and we left together, meeting the police schedule. When we drove into the lot, we were immediately surrounded by squad cars. No guns were drawn. I stepped out of my truck, and one officer stepped out of his car. He walked to me and asked, do you have the pawn ticket? I gave it to him. I reached out with the money. He said, Forget the money, this is on us. The officer walked in, and walked out with Jeff's guitar, placing it firmly in my hands. His lip curled, and his voice cracked as he stepped away.

I missed his words, but they were as real and connected as anything the universe can produce.

All of us experience this sympathetic reaction between people repeated in countless circumstances. We experience that moment when the connection transcends the body politic.

Unintended Intraoperative Awareness and the Transcendent Experience

People with a little water under their bridge usually have a story about being under general anesthesia. Stewart Hameroff has a great deal of experience as an anesthesiologist bringing people in and out of a state, but not sleep. However, not everyone goes under. According to Mayo Clinic (2019),

> Estimates vary, but about 1 or 2 people in every 1,000 may be partially awake during general anesthesia and experience what is called unintended intraoperative awareness. It is even rarer to experience pain, but this can occur as well.

> Because of the muscle relaxants given before surgery, people are unable to move or speak to let doctors know that they are awake or experiencing pain. For some patients, this may cause long-term psychological problems, similar to post-traumatic stress disorder (Overview).

It's possible that unintended intraoperative awareness gets interpreted as an abduction or a near-death experience. Usual confusion, white light, the strange beings looking on, and the odd noises provide a possible mix for this kind of reaction. More common is what doctors and nurses experience with patients in hospice care — people near death who experience dead relatives in complete calm. No agitation. They explain encounters without uncertainty. Dead relatives touch them. Invite them toward a fourth dimension in a state of grace, unencumbered by day-to-day conflict. Taped interviews with these hospice patients demonstrate a complete lack of fear as they describe these encounters. Dr. Christopher Kerr, Chief Medical Officer at The Center for Hospice and Palliative Care, told a TEDx audience in Buffalo, New York that "these universal near-death experiences are medically ignored" (Kerr, 2015).

In nominee Patris et Filii et Spiritus Sancti, Amen.

Among the many issues between hard and soft science, the *ignore the other side* reminds us that experts of every stripe are not much different from career political hacks in Congress. Am I unfair? Well, let's be clear; I don't care. The schism servicing the polemics between hard and soft science is unreasonable. The research community in academia is accordingly enamored with the scientific method, and nothing else will do. The scientific method people tell us that unanswered questions will appear in their tent over time. Sometimes they measure that in hundreds of years. So, ordinary people are supposed to slow-walk this because fundamental equations are our only source? From my perspective, they can have Fundamental Equations, Fundamental Constants, and the Hydrogen Atom, if they understand that 'fundamental' is only a necessary-core-of-central-importance until it changes.

Quantum fields casting for ignition in a particle accelerator do not possess position, momentum, energy, or angular momentum demonstratable any more than boiling water around your poached eggs. So, dying people talking with dead relatives compared to particle physics in quantum wave theory is a push. They both dead-end in guesswork.

Then there are the interstellar science guys who don't care either. Harvard's top astronomer says *an alien ship may be among us — and he doesn't care what his colleagues think* (Selk, 2019, p.1). Early philosophy allowed for this kind of contrarian position. Plato, Aristotle, and company used to chase each other around the Basilica like hammer-throwing gods with preposterous nonsense until some unanimities emerged. But they did not exclude each other's ideas because of perceived truth. They were smart enough to understand that the 'Heavens' were more significant than they were. The Catholic Church did, however, begin to burn people based on its own truth. That turned out to be a monumental mistake. I view these early Holy See decisions as to the idiomatic expression *throwing the baby out with the bath water.* The avoidance of bad becomes the destruction of good.

To all secular scientists, the most annoying guy in the world has got to be Deepak Chopra. And, yet, when one works through The Seven Spiritual Laws of Success, (Chopra, 1994), his blend of physics, philosophy, and other cultural wisdom is not such a lousy soup after all. Especially when we understand that Chopra is willing to take a position outside the wire, as he did with Brian Cox OBE, FRS English physicist, and professor of Particle Physics and Astronomy at the University of Manchester. In this case, Chopra's message was regarding the Cox position on the origin of the universe.

After reviewing Cox's points on Twitter, Chopra sent Cox the following message, "I'm going to shove my cosmic consciousness up your ass" (Cox, 2015). So, I am happy to report that the frailty of the human condition prevails in matters of complex cosmology regardless of a person's attachment to the seraphim of higher orders. In nominee Patris et Filii et Spiritus Sancti, Amen.

We Need to Talk about Mathematics

In space-time geometry, there is mathematics (math). One question being, did we invent math or was it always there. Then the question, where is there? In the 6th century BCE Iron-Age of China, Lao Tzu, the Yin-Yang thinker, thought space harmony was always in place. He was followed in succession by Pythagoras, Confucius, Socrates, Plato, and Aristotle. All these thinkers struggled with method and form as they might relate to an empirical approach to all that was created.

The first woman to enter the mix was Hypatia of Alexandria. Her father was the last, and famous, librarian of Alexandria. She carried his example, as well as Plato's thoughts, forward until she was butchered by a crowd of Christians who didn't like her teaching style. I know this stuff because at a large U.S. university I told my faculty handler he was *an incompetent mindless dystopian place-holder*. There was no fresh firewood available, so they just canceled my contract. Just sayin, if you're worth a shit, style matters!

Finally, the modern philosophers entered the carefully worked-out skirmish: Thomas Aquinas, Niccolò Machiavelli, Francis Bacon, Thomas Hobbs, John Locke, our dog-loving Rene Descartes, Adam Smith, Immanuel Kant, and Edmund Burke. All these people are essential to scientific thought process experimentation, but, to make sense out of order, we needed exactitude. That precision came later and is underway today.

We needed to measure stuff with extreme accuracy. We also needed to look deep into our own brain's corporal reality where there are cells housing molecules made of atoms. The atoms have a central nucleus of protons and neutrons with electrons flying around them—most likely, by recent research, embedded in quantum fields until we look at them and then the field coalesces into a particle. So, the flying around thing is really a graphic metaphor, not a reality. There is mathematics associated with all quantum moments regardless of original discipline. Consciousness in the universe encompasses all we can imagine. If mathematics eludes you, do not despair. Einstein needed help with math if you get my drift.

The Context of First Cause

While advancing further into the mathematics of small things in higher orders, we should have a look at the optics of deists and atheists. According to Pinecrest (2019), the Hon. James Mountain (R), Oklahoma and Chair of the Senate Environmental Committee, said "I am not impressed with science and scientists because the Lord Almighty can overcome all those so-called *facts* in the blink of an eye."

Carl Sagan, the American astronomer, said, "I don't know where I'm going, but I'm on my way" (Wikipedia, 2019).

And C.S. Lewis nails it with "If we find ourselves with a desire that nothing in this world can satisfy, the most probable explanation is that we were made for another world" (Good Reads, 2019).

We quickly see the tension between deists and atheists. As these two groups antagonize each other, it makes the study of quarks and neutrinos of neurons more interesting. Richard Dawkins underwrites his position by letting people know his side, as in we and they, are warm, loving members of society who understand that the universe is going to die a mind-bending death in extreme heat. That the Earth's end is meaningless and empty. Conversely, the Deepak Chopra side believes in spiritual laws leading to the infinite nature of universal consciousness—a metaphysical actuality without end.

Michio Kaku, a string theory theoretical physicist, from the *Maybe a God* side, takes the position that the God question is undecidable and not part of science. He maintains that proving God is like proving the existence of a dragon. He is not alone. Neil de Grasse Tyson is in the same camp in the proof category. Tyson retorts with marked frequency, show me some pictures of God, take notes, ask God about the weather where he is. Michio Kaku leans toward Spinoza and Pythagoras for a

God of harmony and beauty as opposed to the personal God answering prayers and applying vengeance to bad guys.

Michio Kaku steps away from any evidence of God. No need for evidence because in String Theory, which is Kaku's day-to-day adventure, God is cosmic music. And that is the theory of everything as the strings vibrate their way to God. Dawkins finds this stuff "ignorant, stupid or insane" (Berlinski, 2009).

Then we have David Wolpe, the Max Webb Senior Rabbi of Sinai Temple in Los Angeles. Named the most influential rabbi in America by *Newsweek* Magazine and one of the fifty most influential Jews in the world by the Jerusalem Post. He mounts his authenticity argument because he is with dying people as a condition of his job. He knows they are going somewhere else—the dying people.

> That when I hold the hand of someone who's dying, that I really do believe that I usher them into another mode of being. People have believed that since the beginning of time, and the fact that not everyone believes it today doesn't mean it is not true. After all, 'Though he slay me, still though I believe in him'" (Wolpe, 2009) (Does the universe have a purpose or meaning | Michio Kaku vs. Richard Dawkins Debate).

Dawkins continues to hold that all the arguments from the theists are nice, but that doesn't make them true. Finally, Michael Shermer, the absolutely no God guy, manages to say to 682,415 YouTube views "We look for love, career, work, meaningful work, helping other people, and *transcendence*" (Oops!) (Does the universe have a purpose or meaning | Michio Kaku vs. Richard Dawkins Debate).

So, all of them hedge their bets, save, maybe, Dawkins for the atheists and Wolpe for the deists. It all reminds me of the refrain of the world's most famous atheist, Christopher Hitchens. In a conversation with his Jewish Rabbi friend just before Hitchens died, the Rabbi asked him if he was going to accept God before he died? Hitchens reply: "Well, I guess this is no time to piss anybody off."

Window with Quarks and Neutrinos

For us, we are going to keep swimming in the cytoplasm, inside this field and particle window with quarks and neutrinos.

You will remember, as an illustration of scale, this stuff gets small enough that instruments can only report reactions to change about the wave function in the remarkably smaller network of microtubules; hyper small cylinders that dress up like fancy bead-work on a toilet paper roll. As we recall, these guys are assigning fire instructions to neuron axions through neuron gap junctions.

To our probing dismay, this structure independently knows what's going on so we can just put our keys in our mouth and go sit down—we remain flummoxed by how quantum fields work. To get your arms around scale, consider a thought by Fenker (2019): If a nucleus were the size of a basketball, the electron would be flying around two miles away. All the space in between the electron and the basketball-size nucleus is empty (1). So, we are practically empty space.

If you get on a bathroom scale and look at your weight, what you see is a measure of empty space. For equal time, Sean Carroll disagrees with this example, saying that there is no space in between things flying around each other. Everything manifests in quantum fields of infinite density having been that way always and forever. Weight is no more an issue than gravitational enticement of dark matter. This makes it easier to understand our need for multiple tracks of cognitive ability to read through the constructs of quantum physics—upon which Little, Brown and Company, and Candlewick Press used to produce *Where's Waldo?*

You will notice we keep using measure. That means math is a study of its own and is being used to measure our own reality in the platonic world. That would be a measure accurate to 1.673x10 to the high minus 27 kilograms—so small we must enter a different reality independent of ourselves. Now, stand by. None of this is new. It's always been there, this alternate world of approximated sequence mathematics, with its own deviation away from predictability. We measure the natural states of a world view. The 'states' are changing, and we are trying to quantify the change.

The thought process begins in Metaphysics, just as it did for Aristotle. He invented an ontology for cause behind change. But not all things are as they appear. For Aristotle, the Earth was stationary and the center of things.

Today the Earth is moving through space at 67,000 mph (107,000km/h) and not the center of much if anything. Through the centuries, science developed a systematization to find cause. The Earth was stationary, and then it was moving. What is that about? Before instruments, it would have been a philosophical tradition. Aristotle went for cause, and Spinoza stacked on 'Sufficient Reason' cause not being enough. Cause needed a reason to make a move (Stanford University Library, 2019). We understand things do not just happen; however, in quantum mechanics, that is the case. Things in Planck Scale geometry just happen.

In math, there are axioms that originated as artifacts of our civilization. Before scientists measured fire sequencing in Axons, the Babylonians used arithmetic in commerce—also, geometry. But, as we lifted our eyes off the horizon, we began to think of the universe as a formal system. This formality caused us to think in mathematical theorems. We began making statements around general structures that were self-evident. Aristotle admonished that we could not communicate science if the exchange was not built around the truth of Axioms (Barnes& Griffin, 1999).

There had to be beauty in the form of mathematics. Balance if you will. 2 + 1 = 1 + 2. We worked to see this design balance throughout the universe. Over time, these designs were solvable—some were unsolvable. As humans, we tend toward the path of least resistance. So, logic tells us we focus on the math solvable. We avoid math unsolvable. Now we call for the question: In all that exists, how much of human math is relevant to all the possibilities?

Consider, as we go forward, we are structuring discovery by looking back. Our artifacts are supporting scientific progress. In some quarters, this backward-looking progress is correct and non-negotiable. Therefore, is forward exploring profoundly important, or not?

In quantum mechanics, we give up the artifact of causality. As fundamental as causality is, we cannot justify a cause in the lowest reaches of wave-particle theory. We now call this a new level of reality. A convenient turn-of-phrase to explain we have a situation that does not match any of the laws of physics. Sean Carroll gives a comparison of us to a new reality in "The Big Picture." He likens us to Wile E. Coyote. That moment when he runs off the solid edge of the cliff and falls when he notices he is in thin air. Carroll then points out a hazard of artifact tradition.

Our scientific vocabulary is lagging the actualities of quantum mechanics (Carroll, 2016).

Mathematics is not an Explanation

Nothing in the universe / multiverse / road warrior movie, and so on, is the way it is because of mathematics. Mathematics is the way it is because of the matter we see or experience. Physics drives math, not the other way around. Metaphysics is how the ideas emerge. We forget that when we began lifting our eyes off the horizon, we were not looking through a Bushnell 20mm range finder. We were only looking up in wonder. The math came later as we subscribed shape—the cube all the way to the icosahedron while looking for symmetry that appeared beautiful. Some were better at this than others.

Paul Dirac: The man who conjured laws of nature from pure thought—A fellow quantum physicist said his discoveries were like 'exquisitely carved statues falling out of the sky, one after another' . . . one of the most baffling geniuses the world has seen ("*The Guardian*," 2019).

In Merleau-Ponte (1996) the Phenomenology of Perception pursued how one defined and understood reality. In Merleau-Ponte's perception of how to see things, the perception is modified by the actuality. Remember our Évariste Galois who managed to get himself in a duel over Stéphanie Potterin du Motel? Just before he took the pistol's ball in his head, he revolutionized mathematics by showing in mathematical summitry—group theory—things have an appearance which doesn't change when the point of view is changed, Juxtaposed to Merleau-Ponte. This is the difference between Existentialism and Mathematics. The math guys seek ways to fix things in space-time geometry. The metaphysics guys get all fidgety when stuff doesn't change. Hard scientists say the constants (laws) of nature. Soft science guys say the habits of nature. Mathematics needs to nail down material so it can be reproduced repeatedly. Metaphysics sees no progress unless the horizon is continuously changing.

I think there is room for both. They both use trial and error. Whether they admit it, both operate in the day-to-day world of pedestrians. Planes fly because of pressure differences around shapes not symmetrical. Wars are intoned using Plasma Physics. Mathematical design can be lovely, attractive, gorgeous, or magnificent. Which word describes the chaotic inflation of 500,000 tons of TNT?

Everything you can imagine exists

The most exciting developments in measurable science are the media optics suggesting we might be able to teleport something more significant than an individual particle. Or, the talk of multiple universes. In the measurement domain, we see 10 billion light years in all directions. The new idea is we might teleport to a parallel universe. That superposition universe we know is there, but we can't see it. For example, put a crystal prism on your windowsill in the sun. We live in the visible light tossed on your ceiling. Our no-see-em parallel universes could be on both sides of the visible spectrum: Radio Waves, Micro Waves, Infrared Light, UV Light, X Rays, or Gamma Rays. We live in all this tackle at the same time, just in different no-see-em places on the spectrum.

So, maybe your crystal is the bellwether for superposition of multiple universes. For the skeptic, ponder "Principle of Fecundity" by Robert Nozick—holder of the Joseph Pellegrino Professorship at Harvard University? He said *Everything you can imagine exists. All possible worlds are real.*

These different universes could come in different measures. Ours is supposed to be caused by chaotic inflation (Big Bang). That theory is changing. Andrei Linde at Stanford University thinks chaotic inflation was cold until it got hot. Then a bubble formed and produced our Big Bang. The difference now is the universe is going through this process at high frequency with chaotic inflation happening, perhaps, millions, billions and trillions of times.

Speed is important here. I heard an interesting comment the other day. The speed of light is 186,000 miles per second, but space can go as fast as it wants. That kind of plays into Linde's theory. There cannot be arbitrary assumptions. Hard pill to swallow for traditional researchers. The deeper we get into platonic reality, and its relation to chaotic inflation, the willowier becomes the theory.

And that is underwritten by Pythagoras who plucked and sang his way to harmony. He established a relationship between object and sound. For Pythagoras, it was the hand of God on the violin scroll that tuned the universe.

The Crow

In "God the Father, God the Son, and God the Holy Spirit" The Dove Might Be a Crow

I was watching Sean Carroll, at the Royal Institution in London, explain the search and finding of the Higgs Boson. He put the subject in perspective by reminding us of the significance of the Higgs Boson; is nested in wave theory, not particle theory. The Higgs produces a bump in the wave. That bump is the gathering of mass that was heretofore unfounded. What that simple find means is that all that we have growing in the universe, including ourselves, is dependent on the Higgs Boson doing its mass attracting trick. I was not brought to tears as I heard the news. However, it's easy to understand how the scientists trying to find it for the past forty-five years would be, and they were. Next time someone mentions Higgs Boson, look quietly at your hands, then the sky, and think where you are in space-time geometry.

The Higgs Boson contribution to consciousness—not just ours—advances unilaterally throughout the cosmos.

Scientists systematically study the behaviors and brain states of Earth's creatures. Now each year brings a raft of new research papers, which, taken together, suggest that a great many animals are conscious.

It was likely more than half a billion years ago that some sea-floor arms race between predator and prey roused Earth's first conscious animal. That moment, when the first mind winked into being, was a cosmic event, opening possibilities not previously contained in nature.

There now appears to exist, alongside the human world, a whole universe of rich animal experience. (So much for René Descartes, however, I think my shape-shifted Wolf picked him off earlier in the text) Scientists deserve credit for illuminating, if only partially, this new dimension of our reality. But they can't tell us how to do right by the trillions of minds with which we share the Earth's surface. That's a philosophical problem, and like most philosophical problems, it will be with us for a long time to come.

Apart from Pythagoras and a few others, ancient Western philosophers did not hand down a rich tradition of thinking about animal consciousness. But Eastern thinkers have long been haunted by its implications—especially the Jains, who have taken animal consciousness seriously as a moral matter for nearly 3,000 years (Anderson, 2019, para. 1).

At the University of Washington in Seattle is the nationally acclaimed Avian Conservation Lab. Here academics focus on crows like 12,000 at a time. Then only one at a time. Sometimes two at a time. What they know is a Crow recognizes a human face. Crows operate just like us. They commute to work. Crows in specific neighborhoods return to those places every day—every Crow workday. At day's end, they fly home to roost outside the city. In the early morning, they come back to Crow work at the same place—every day. So, when you wake up in the morning and see a Crow on the powerline flanking your house, that Crow is at work. When the Crow looks at your face in the window, he or she turns to a partner Crow and says, Yep, just like yesterday, Sapien-lazy still hasn't made the coffee.

Then they go about their work detail. Typically, that would be cleaning out your gutters as they look for grubs. They use sticks to open small areas. If they get a windfall slice of food from a passerby, one of them will take charge of the new food and carry it off to a fake location to deflect the other from stealing it away. Then, in the cover of activity, move it to a second location, so the new food is isolated to the knowledge of that Crow. The other gets to search.

If the Crows feel threatened by an intruder, they will raise all kinds of noise, fly around, dive bomb, and squawk at volume-10. They will fly within inches of the target. If it's a large bird, they will team up and go for the wing—pulling out wing feathers until the invader is unable to fly. While this is happening, they are calling for backup from their fellows in the neighborhood. And, the support team will arrive en masse. When all is over, they fly in circles, squawk in unison, and celebrate their success. Cosmic Consciousness in birds is scientifically suspect among some inquirers. When they die, nothing happens. They are just dead and rotting away to the grubs and ants that find them. Others see birds possessing neurological substrates.

My Dead Crow Experience

I made a mistake one day of shooting a Crow. I had a BB pistol and saw this Crow high in an evergreen tree. I made this mindless gesture of aiming at the Crow and squeezed off a shot. The Crow jumped off the limb and began a spiral descent to the ground. It was like a Second World War fighter plane that was clipped on one wing spinning toward the ocean.

Before the Crow hit the ground, I could hear the noise. Within seconds there were dozens of Crows in the air around the tree. Then one of them came for me. I dropped behind a stair banister. I turned toward the tree. The cacophony of sound, mixed with endless black wings, scared the hell out of me. I kept returning to Alfred Hitchcock. What did Hitchcock know about birds that we didn't?

I felt guilty. I gathered myself and walked toward the base of the tree covering my head in a gray hoodie. The closer I got the more the excitement among the birds. The din became deafening. The anger was everywhere. If they could kill me, they would have. I spoke out loud and tried to apologize. They weren't interested. The Crow was dead. I was empty. And the Murder of Crows was outraged.

I thought it would be an acceptable gesture to bury the Crow in my yard. That was my second big mistake of the afternoon. I walked away with my life. I sat on my back steps and watched the Crows circle, eventually landing in the tree en masse. At twilight, they left. They flew, as they always do, home for the night. When the sky was clear, I walked back toward the tree to recover the Crow and apologize to the universe. The Crow was gone!

Mark Twain on Crows

> I suppose he is the hardest lot that wears feathers. Yes, and the cheerfullest, and the best satisfied with himself. He never arrived at what he is by any careless process, or any sudden one; he is a

> work of art, and "art is long"; he is the product of immemorial ages, and deep calculation; one can't make a bird like that in a day. He has been reincarnated more times than Shiva; and he has kept a sample of each incarnation and fused it into his constitution. In the course of his evolutionary promotions, his sublime march toward ultimate perfection, he has been a gambler, a low comedian, a dissolute priest, a fussy woman, a blackguard, a scoffer, a liar, a thief, a spy, an informer, a trading politician, a swindler, a professional hypocrite, a patriot for cash, a reformer, a lecturer, a lawyer, a conspirator, a rebel, a royalist, a Democrat, a practicer and propagator of irreverence, a meddler, an intruder, a busybody, an infidel, and a wallower in sin for the mere love if it. The strange result, the incredible result, of this patient accumulation of all damnable traits is, that he does not know what care is, he does not know what sorrow is, he does not know what remorse is, his life is one long thundering ecstasy of happiness, and he will go to his death untroubled, knowing that he will soon turn up again as an author or something, and be even more intolerable capable and comfortable than ever he was before.—*Following the Equator* ("Twain Quotes," 1898).

Crows' flight speed is regularly at thirty to sixty miles per hour with speeds in excess of seventy miles per hour while diving on predators. Crows roost in groups up to two million birds. Crows use more than thirty distinct calls. Their color vision is well-developed, and they see polarized and ultraviolet light.

"U.S. Army Major Charles Bendire, one of America's first ornithologists, who describes a pet crow named Jim that was teased with a knife. The knife was often brandished just out of Jim's grasp. Insightfully, Jim bit the teaser's hand, not the knife, causing the knife to be dropped. Jim picked up the knife and flew it away to a hiding place Bendire never discovered" (Marzluff & Angel, 2005, p. 47).

The way a crow
Shook down on me
The dust of snow
From a hemlock tree

Has given my heart
A change of mood
And saved some part
Of a day I had rued
Robert Frost, "Dust of Snow, "1923

After the Crows

As I look at that title, After the Crows, it reads like a Bible Chapter from the Plagues in Exodus. I began to think about what a Native American friend of mine referred to as the river of life. In fact, that's the name of his book, *The River of Life* (Marchand, 2013).

Native Americans (I prefer Indians) have honored the trillions of creatures around us through Indian spirituality for seven generations that we can codify in art, and 12,000 years in bushcraft. Indian spirituality is based on we are-one-with-nature, or one with "Haweyenchuten" (Wenatchee for Creator-God). As Europeans et al. focus on stars and galaxies, Indians focus on Earth as a living system with stars and galaxies.

There is also the Greek Goddess of Earth, "Gaia" (Sarewitz, 2015). There is plenty of Earth associated with God to go around. Also, it's not new. Right after the Ice Age—about 12,000 +/- years ago—when we entered the current interglacial period of Holocene, we began honoring our companion animals in art and craft. Beliefs came forward that if animals were not honored in death, man, the hunter, would be forthwith unlucky in the hunt. Of course, 12,000 years ago, being the hunter or hunted was a push; however, man's evolving through language turned out to be important.

On the other hand, Ross Anderson wrote this in *The Atlantic* regarding the "Neminath" to devotees of Jainism, "Many people will come here from all corners of the Earth to pay their respects to Neminath, who is, after all, only a stand-in for whomever it was who first heard animal screams and understood their meaning" (Andersen, 2019, p. 30).

By now, everyone reading this book knows my position on the relationship between man and animals, and my position on René Descartes, whom I find disgusting, his contribution to science notwithstanding. Having said that, let's move on.

In *The River of Life*, Mike Marchand, along with contributions from fifteen members of the Confederated Tribes of the Colville Reservation, developed a cultural comparison between Western culture and Indian culture. It was a root cause comparison, comparing brain hemisphere traits with Indian culture being Right Brain dominate.

- For Example
  - On the right-brain side, we find perception, creativity, art, intuition, spiritual, music, dance, imagination, and humor.
  - On the left-brain side, we find order, logic, math, details, science, reading, facts, writing, planning, systematic, linear, and goals.
- As these are compared between cultures,
  - Western culture becomes competitive, aggressive, controlling, with time orientation.
  - Indian culture becomes cooperative, emotional, orientation to present, control of self, and participation after observation.
- When reduced to a value system,
  - Western culture indicates: Only humans have spirit; nature is to be conquered and teamed; scientific methods used for healing; the more you own the better you are.
  - Indian culture indicates everything in nature has the unique spirit; humans should live in harmony with nature; healing by being in harmony with nature; sharing in generosity is greatly valued.

Here is the tension. The Indian culture traits were based on perceptions of original tribal people. None of the researchers took this forward to today, because now all they find are tendencies, not actualities. Indians have been integrated into the dominant society.

Marchant et al. go on: "Some consider Indians to be the keepers of the Earth or as aboriginal times keepers of the fire . . . this connection lies beyond both psyche and physics . . . [many] scientists have concluded that the universe is a living organism and that we are all connected at the quantum level" (Marchand, 2013 p. 87). So, as with Hameroff and Penrose, Indians agree, there is something else.

The Coyote

A few years ago, I was working with the Colville on a joint education program to open a branch campus on the Colville Reservation for Benedictine University in Chicago. That initiative was connected to a Benedictine University and China student exchange, wherein students of Benedictine and the Colville would exchange for the last two years of a combined Baccalaureate degree from Benedictine University and a sister school in China. We also held talks with faculty members of The George Washington University at Foggy Bottom. Neither program materialized for the Colville. However, my relationship with the tribe took an animating turn concerning their relationship with the Earth.

I first met Mike Marchand at an Indian casino where he was attending a Northwest Tribal Conference. I didn't know much about Indians. I was supposed to have a conversation with him about a possible Benedictine branch campus with a Ben-U exchange in China. On schedule, I walked into the casino hotel lobby. I felt like I was on a movie set. I had seen all these clothes in the movies.

There was a woman standing near me. She seemed comfortable, so I approached her to see if she knew [a] Mike Marchand? She looked at me like I was from another planet—and I was—and pointed to this short white-haired guy across the lobby and in a banquet room. He was wearing Indian like clothes, sort of understated,,,,but still culture-specific. He was expecting me, so all I had to do was attract his attention. I waved my hand like one would expect to see a wimpy understudy wave at the director. But I got lucky, and he walked toward me. The Indian woman who identified him let me know, I am imaging for my fish-out-of-water look, "He is the president of the organization—Nineteen Tribes in all. He is important and powerful."

Well, shit, I thought. *Why couldn't I start off my Indian conversations with someone like her, instead of a powerful warrior politician? This guy will probably want me to sit in some hot rocks tent and speak in an altered flow about disavowing the white man's ways.*

As he got closer, he began to resemble El Chapo with Marlon Brando's face. He was a shortened version of the Godfather. He moved the same way, slow with his head swivel in constant proximity awareness. He extended his hand motioning me to sit across from him at a lobby table. Still, no words were spoken. I addressed my chair while looking carefully at him as I sat down. Still, no words were spoken. No handshake. No talk. We sat looking at each other.

Here I could write, "It seemed like forever." Not the case. It was longer than forever. This guy was looking at the most uneasy, culturally inept, over his depth, white man he had ever seen. Still no talk.

His eyes moved away from mine. He looked at the woman, still close to me but out of my view. His eyes came back to mine. Slowly, he leaned in and, with his head cocked to one side, whispered, "I have a Bachelor of Arts in Economics and Urban and Regional Planning, a Master MS NSF Fellowship for Bioenergy, and a Ph.D. in Forestry from the University of Washington.

"I speak English."

One of the sites proposed belonged to an investment company from Japan, now operating in Washington State. The Colville Reservation is in north-central Washington, close to the Japanese organization's property. As a group, several members of the Colville Tribe and I went to see the site. As we drove around fifty million dollars worth of sagebrush and rocks fronting on recreational lake frontage, we came over a rise—a panorama view to the horizon in every direction. We were stopped by a rock circle about fifty feet in diameter and five feet high. We got out of our trucks and walked to the circle. One of the men in our group was an archeologist. He walked into the circle, bent over and put his nose right against the lichen embedded in the rocks. He said, "I think this circle has been here for a long time. I think we should take a sample." So, we did.

About a week later, I got a call. The rock lichen is 5000 years old. Then I found out the National Association for Stock Car Auto Racing (NASCAR) people had been to the site as a possible new track. And, because part of the land fronted close to federal Interstate Highway 90, a Casino was in the plan. And, of course, a gazillion other things associated with significant developments. But the lichen was 5000 years old. And the circle, well, that was an ancient prayer circle according to the archeologist.

We drove back to the site, and to the circle. The Indian in my truck with me was an old Rock & Roll junkie who did some stand-ins with Red Bone. He was way cool in his parlance. He decided to come back, sit in the circle and pray. I am by the hour trying to aggregate what's happening. We have a University, a Casino, NASCAR, and developers hanging over the fence, all wanting a piece of the deal.

Mike Marchand decided to step into the mix, and he offered to have the Confederated Tribes of the Colville Reservation buy the entire property for thirty-six million dollars. The owner turned down the offer and continued toward the elaborate plan. I go home depressed.

Now, in Colville Indian culture, the Coyote is the image of Indian Spirituality. And the Earth is a shared domain—making that clear by reminding us that in every glass of water we drink, some of the water has already passed through fishes, trees, bacteria, worms in the soil, and Coyotes ("Water," 2019).

We are back at the site. There is tension all around. The Indians are considering upping the offer. NASCAR was in Seattle with lawyers. The Benedictine University president was calling me from China where he traveled to make deals with other schools. My Red Bone Indian buddy was hungover. The lawyer for the property owner drives up in a red SUV. He steps out of his ride. Walks toward us. He was known to be campaigning for the NASCAR deal plus a massive development. Everyone is silent. I can hear a fresh williwaw wind start to shudder the surface of Moses Lake. I can smell sagebrush. I am in a kind of superposition with this place and some other place. I don't know where.

Then, as if we are a dynamic flock of starlings, we all turn in unison, and there it was—a Coyote in the middle of the stone circle. I didn't know it at the time, and neither did the lawyer for the owner and NASCAR, but the deal was off . . . there would be nothing built on this space. At that moment, we were standing on a 5000-year-old sacred site. The Coyote was now the hand of Mother Earth, and the Law was on her side.

Mother Earth stepped up and took the deal for herself. It was a wise choice. Outside lawyers got involved. I met with my own stark reality of who they really were. I stood, told them, and the associated tribe elders, the developers were carpetbaggers working at Indian expense. Today, I hold a ceremonial Chief Blanket from The Confederated Tribes of the Colville Reservation.

Writing in Scriptio Continua

Native American (Indians) keep the spirit's fire. However, there was a time before our time when no gods of any kind became fashionable. It was about December of 1417 when the lost manuscript "De rerum natura" by Roman poet and philosopher Titus Lucretius Carus, reappeared in a German monastery, found by a Renaissance book hunter, Poggio Bracciolini.

De rerum natura ("On the Nature of Things" or "On the Nature of the Universe") transmits the ideas of Epicureanism, which includes atomism and psychology. Lucretius was the first writer to introduce Roman readers to Epicurean philosophy. The text, written in some 7,400 dactylic hexameters, is divided into six untitled books and explores Epicurean physics through richly poetic language and metaphors . . . the universe described in the poem operates according to . . . fortuna, "chance," and not the divine intervention of the traditional Roman deities (Wikipedia, 2019).

The writing of Lucretius evolved into the Renaissance, changing how we think. It changed discovery toward deconstructionism. The leverage of religious power would swing to secular power. The De rerum natura influenced the trifecta of Freud, Darwin, and Einstein. In the United States Declaration of Independence, Thomas Jefferson annexed its postulations.

The Deep State of the Middle Ages was fading to black. The intrigue of the transition is best found in *The Swerve, How the World Became Modern*, a benchmark codex by Stephen Greenblatt, Professor of Humanities (among other things) at Harvard (Greenblatt, 2008).

"The advantages of tossing out compliance was the primacy of bodily pleasure, the advantages of moderation, the perverse unnaturalness of sexual abstinence, no rewards for the dead, certainly no punishments either" (Greenblatt, 2008 p. 223).

The inevitable tension between the reach of the Steller and the grasp of the Interstellar had begun. As in all transitions, the curious nature of Sapiens prevailed. De rerum natura was issued to the public with warnings of its absurdity. That we could be unsupported and made of random small particles circling each other was laughable. Without the hand of God, and more importantly, the Holy Roman Empire guiding our existence, we were doomed to extension by gambling with this secular freedom of the human condition. The dialogical disavowal was underwriting both sides of the argument.

The switch began to bleed changes into Philosophy, Science and Invention, world events, and Culture. As our purpose in life changed to finding happiness, the world philosophers took to the secular mechanism. The Greeks with Epicurus's move to satisfy the right desires; the Roman Epictetus's aggregation of power, knowledge, and opinion; the six classical systems of Hindu philosophy; and St. Augustine of Hippo deciding man can choose between good and evil, heretofore unnavigable without the hand of God.

In science and invention, Archimedes develops hydrostatics (objects that sink or float). Hipparchus divides the night sky into longitude and latitude and shows the world in precession. Hero of Alexandria comes with the first steam engine, and Ptolemy the Greek decides that the sun revolves around the Earth—close but no cigar.

In world events, China Qin Shi Huangdi manufactures his Terracotta Army—the first notion of a production line as the clay soldiers moved through their construction. Julius Caesar gerrymanders the Julian calendar. Jesus Christ is crucified. Vesuvius erupts doing away with Pompeii and Herculaneum. And, finally, in concert with nationalism, the Great Wall of China begins its march. The Rosetta Stone is carved in Egypt; the Nazca Plan drawings evolve in Peru, and the age of Maya begins in South America.

But spiritualism was not going to go down easy.

In c.1077 St. Anselm brings the ontological argument that imagining a godlike figure means that the likeness must exist in space and time. To focus on the image is proof of God. Bodhidharma thinks up Zen Buddhism at the Shaolin Temple in China. Japanese culture takes a turn toward Shingon Buddhism. Anicius Severinus Manlius Boethius, born into the Roman aristocracy, takes a page from his powerful socialization and decides to move God outside of *time* so God can see past, present, and future.

Now, with God riding on the back of recovery, enter the Prophet Muhammad and Islam. William the Conqueror leads the Norman Invasion. Westminster Abbey is consecrated in London. And, for pilgrims by the millions forever more, the cathedral of St. James at Santiago de Compostela is built in Spain.

Working conversely against the Santiago de Compostela pilgrims, mechanism begins to surface inside France. Michael de Montaigne became the first official skeptic (skepticism) in 1580. Francis Bacon designs the scientific method. And, Frenchman René Descartes kills his wife's dog while uttering "I think, therefore I am" (Cogito, ergo sum). The scientific method jumps into the queue. Isaac Newton starts looking through a telescope. Italian physician Santorio Santorio uses a thermotic process to measure human temperature. The Royal Society is founded in London. And Robert Boyle decides the volume occupied by a gas is inversely proportional to its pressure.

With all this new stuff on the shelf, thinkers and arguers began to codify between ce1690 and ce1799. We are now modern. John Locke begins with Tabula Rasa (Mind as an empty slate) knowledge comes through experience. Isaac Newton presents his Law of Gravity. David Hume decided the philosophers were wrong, and we are not so smart after all. Then Benjamin Franklin flies a kite, and David Hume reconsiders.

Secularism is now on the march—Adam Smith publishes "The Wealth of Nations." America declares itself away from England. Dollars are introduced as US Currency. James Watt tweaks his steam engine. And as a simulcast, France sees the execution of Louis XVI and Marie Antoinette, as George Washington becomes the first President of the United States.

From here on in thinkers think about thinking AKA Aristotle. We see Karl Marx, John Stewart Mill, Fredrich Nietzsche, William James, George Santayana, John Dewey, Edmund Husserl, Martin Heidegger, Bertrand Russell, Jean-Paul Sartre, John Rawls, and finally, we notice a woman with Simone de Beauvoir.

The Existentialists

With 2000 years of growth under our belt, we needed perspective. As a community of science exploration, we were becoming familiar with the notion of empty space in the cosmos. Historically, philosophers could be mechanistic about what they saw. We see what we see, and work around that. But now we are in the second great war. In the first war, the cosmos was absolute. Now, in this second war, the war machine extends into space. We fly high enough to see the sky get dark in daylight. Weapons visit upon us from unseen space at astonishing speeds. Political reactions are based on fear instead of ideology. Decisions are made to kill people we know nothing about. The other culture isn't making a move on our home ground. We are in the war mix because it's the right thing to do. We needed to think about that. We needed to begin seeing things as they are.

That notion became the Theory of Perception. In the first war, content and object were the same. In this new war, we saw a content separation through hallucination. Ships and planes matched the sea and sky. Some weapons were representations of weapons. Cutouts like duck decoys. We saw the same content; except they were not. The Second World War got us thinking in abstractions. These abstractions were being underwritten by the Existentialists of pre-war and eventually post-war Europe.

Martin Heidegger challenged the meaning of Being (Periñán 2018). The predicate for existence became ontological. If an inference is present, the inference becomes part of existence—even if the inference is for non-existence. Philosophers began to chase their tails around vernacular. These thought experiments eventually found their way to God.

Café Society in Paris Sweeps God Under the Carpet

Not only does God play dice, but he also plays with nonlocal dice in the multiverse. Philosophy as a discipline could not have imagined the reality of dead and alive at the same time. That was too over the top.

> Consider the analysis of truth-condition of "God does not exist": there is something in the world that is picked out by the name God and that thing satisfies the description "does not exist." For "God does not exist" to be true, then, God would have to both exist (in order to be picked out by the name God) and not exist (in order to satisfy the description "does not exist"). Clearly, this could never be the case; nothing can both exist and not exist (Reicher 2016).

Quantum physics made that conclusion false.

Postwar intellectuals who were terrified by the prospects of infinite space and tended to hide in the bohemian intensification of the café society in Paris. Phenomenology was moving toward Paris, from Berlin, as abstract axioms and theories in a human study. Quick to consider phenomenology as an alternative to philosophy were Simone de Beauvoir and Jean-Paul Sartre. If it hadn't been for Beauvoir and Sartre, as inquisitors at Bec-de-Gaz on the rue du Montparnasse in Paris, this new dimension might never have marshaled the interest of the intellectual community. But it did.

The burgeoning phenomenologists champion in Europe was Edmund Husserl, a German philosopher who established the school of phenomenology using his vision of foundational science based on the so-called phenomenological reduction. The argument being, transcendental consciousness establishes the parameters of all possible knowledge. As the phenomenology school gained traction, so did the café society in Paris. The players were young, between ages 20 and 30, intrigued with the excitement of international impact, and in concert with reducing observations to their basic intentionality, loved to make phenomenological reductions of how a woman's breasts pool on her chest as she reclines to 90 degrees. *Who wouldn't enjoy that?*

Free thinking was available in the streets of Paris. Young intellectuals were leaving Germany ahead of the Nazi expansion. The Catholic church was having a hard time getting out ahead of Hitler. The Jews were hiding anywhere they could. The Germanic Polish population, as dominate Catholics, felt they were being compromised, and they were. There was little reason for European youth to find credibility in official infostructure. God did not appear to be the intervener in charge of the established moral authority.

These were some of the reasons why most Christians in Germany welcomed the rise of Nazism in 1933. They were also persuaded by the statement on "positive Christianity" in Article 24 of the 1920 Nazi Party Platform, which read:

> We demand the freedom of all religious confessions in the state, insofar as they do not jeopardize the state's existence or conflict with the manners and moral sentiments of the Germanic race. The Party as such upholds the point of view of a positive Christianity without tying itself concessionally to any one confession. It combats the Jewish-materialistic spirit at home and abroad and is convinced that a permanent recovery of our people can only be achieved from within based on the common good—before individual good (Koehne, 2018).

As the virus spread, Edmond Husserl connected with Martin Heidegger, another German thinker/philosopher who unfortunately compounded the Nazi Party Platform problem by falling ass over teakettle for Adolph Hitler's Nationalsozialismus. That changed as time passed, however, while in the Nazi light, Heidegger joined the Nazi Party and sported the Adolph mustache under his nose.

Heidegger is best known for his contributions to phenomenology and existentialism, though as the Stanford Encyclopedia of Philosophy cautions, "his thinking should be identified as part of such philosophical movements only with extreme care and qualification" (Rothman, 2014).

> The sans-spiritualism problem was also pervasive in modern English literature. Ezra Pound, an outright fascist and anti-Semite who is nonetheless recognized as one of our greatest poets, and T.S. Eliot, also an anti-Semite, are prime examples. Martin Heidegger, a Nazi party member and collaborator, Benedetto

> Croce, who at first welcomed Mussolini's fascism, and Paul de Man, who wrote dozens of pro-Nazi and anti-Semitic articles in the early 1940s, are considered some of the most influential philosophers of the 20th century. Leni Riefenstahl, a giant in the history of film, was an unrepentant Nazi to her dying day (Wheeler, 2018).

This was the backdrop for the beginning of Existentialism and the removal of God from the vernacular. Existentialism reach includes Freethinkers, Secular Humanists, Anti-theists, Skeptics, Rationalists, and Brights, among others.

How it Shakes Out Today

Media optics in the land of God/No God

The protagonists of this fight are outspoken in their zeal for answers. On the No God side, Richard Dawkins is among the principal players.

> "A lively writer . . . an entertaining read . . . Dawkin's outrage at the persistence of medieval ideas in the modern era is warranted. In fact, it's overdue."—*The Nation*

> Lots of good, hard-hitting stuff about the imbecilities of religious fanatics and frauds of all stripes."—*New York Times Book Review*

> "Dawkins is frequently dismissed as a bully, but he is only putting theological doctrines to the same kind of scrutiny that any scientific theory must withstand."—*Scientific American*

Answering Dawkins is David Berlinski. A senior fellow at the Discovery Institute in Seattle. Of his writing we find:

> "An incendiary and uproarious work of learned polemical writing, unique in its scientific sophistication and authority. Rather than criticizing science from the outside, Berlinski excoriates its atheist pretensions from within."—*National Review*

> "In a brief review, it's impossible to convey not just the complexity of some of these issues, but the clarity, keenness, and breadth

> of learning that Berlinski brings to bear on them."—*American Spectator*

> "With high style and lighted-hearted disdain, David Berlinski deflates the intellectual pretensions of the scientific atheist crowd. Maybe they can recite the Periodic Table by heart, but the secular Berlinski shows that this doesn't get them very far in reasoning about much weightier matters."—Michael Behe, bestselling author of Darwin's Black Box and the Edge of Evolution

Dawkins writes of the God Delusion, and Berlinski writes of the Devil's Delusion. Both writers are well-educated. Both writers are acknowledged in the public domain.

Threading the needle between Dawkins and Berlinski is Rupert Sheldrake. Of his writing we find:

> "Rupert Sheldrake is one of the most innovative and visionary scientists of our times. Rupert will be both vilified and praised for his theory of morphic resonance. Whatever your opinion of his work, he will not be ignored. In my opinion, his contributions will be recognized one day on the same level as those of Newton and Darwin."—Deepak Chopra

> "Books of this importance and elegance come along rarely. Those who read this new edition of *A New Science of Life* may do so with the satisfaction of seeing science history in the making."—Larry Dossey, M.D., author of *Recovering the Soul and Reinventing Medicine*

> Science is often portrayed as a paragon of intellectual freedom. It's a quaint idea, but it's not true. Some key concepts in science have hardened into unshakeable, unquestioned dogma. *Science Set Free* exposes ten of the key dogmas of modern times. If even one is slightly off, then the scientific world is in for a shock, and the aftershocks will have huge impacts on technology, medicine, and religion. Rupert Sheldrake skillfully examines each dogma

and argues, with evidence, that all ten dogmas are wrong. After reading this book, I am persuaded that he's right. If you agree that science must be freed from the shackles of antiquated beliefs, then read this book. If you don't agree, then read it twice (1).
—Dean Radin, Ph.D., author of *The Conscious Universe*

The Foundation of Summa Theologica

If Pythagorean harmony and choreographed biology are intended to be graceful, the arguments inside the wire of theism and atheism are tumultuous. The constant anxiety around the "tough question" leads to their own version of chaotic inflation. The disciples of theism and atheism continue pushing dogma. It might be that, despite his flaws, German Existentialist Martin Heidegger got it right. In his writing *Being and Time* he comes with the yet to be answered single question, "Why is it that humans are the only beings that wonder how it is that they exist" (Periñán, 2018).

And that wonder became exclusive to man long before Heidegger. Yuval Noah Harari writing in Sapiens (Harari, 2015) reminds us that any number of species could have come up the chain as we did—but didn't. We could be lions. We could be condors. What we are, however, is a brain capable of abstraction. Representing the two sides: Dawkins makes a case for a selfish gene: Chopra gives it to divine intervention. Harari gives us a refreshingly frank answer, "We don't know."

Matthew Fox and Rupert Sheldrake made their case for the universe being organic writing in *The Physics of Angels: Exploring the Realm Where Science and Spirit Mee*t (Fox and Sheldrake, 1996). They compare the organic structure of all things small to all things big. If one takes his or her experience with visual comparisons, the organic hypothesis of sameness in big and small becomes as good as any other. For example, go to a search engine like Google, and cast around for pictures of outer space and microscopic space. They look remarkably similar. And, this is not lost on science writer Nicholas Bakalar at *The New York Times*.

> The method for measuring the number of galaxies in the universe is, by comparison, straightforward, but it is not easy. Astronomers pointed the Hubble telescope at a portion of the sky of known size, counted the number of galaxies they could see, and then multiplied to estimate the number of galaxies in the observable universe.
>
> There are, of course, complications. Galaxies merge over time; the universe expands; the distance we can see with improved technologies increases. And of course, we are talking only about the observable universe—the part we can see.
>
> But the best estimate now is that there are between 100 billion and 200 billion galaxies in the universe.
>
> So, more galaxies or more cells? This is not a close call. Even using the highest estimate for galaxies (200 billion) and the lowest estimate for human cells (1 trillion), there are at least 800 billion more cells in your body than there are galaxies in the known universe (Bakalar, 2015, para. *Science*).

Once again, these similarities tend to make a beginning case for Panpsychism. Further, scientists in biology and astrophysics seem focused on finding these organics as outer space actualities, not simply visualizations. So, the organic structure of all that exists is believable enough to catch the attention and financial consideration of hard science. Fox and Sheldrake go on to discuss electrometric patterns brought to us by our (star) sun, as well as how cosmic intelligence appears among the writings of Dionysius the Areopagite, Hildegard of Bingen, and St. Thomas Aquinas in *Summa Theologica* (Aquinas, 1485).

The Management Team in Summa

Summa Theologica comes as the splinter Aquinas hallmark in his twenty-six volumes resembling a set by Encyclopedia Britannica. Throughout, Aquinas attempts to justify then current cosmology with Augustinian and Aristotelian traditional thought. In this effort he uses angels. To his sophistication and credit, he employs beings which are recognized throughout the world; not least, the universe. Aquinas gave credit to God for the creation. However, he added-in middle management and front-line employees in the form of angels to manage the process. As it turns out, the angels in the Aquinas mind were most like (light) photons.

According to Wikipedia (2019),

> An angel is generally a supernatural being found in various religions and mythologies. In Abrahamic religions, angels are often depicted as benevolent celestial beings who act as intermediaries between God or Heaven and humanity. Other roles of angels include protecting and guiding human beings and carrying out God's tasks. Within Abrahamic religions, angels are often organized into hierarchies, although such rankings may vary between sects in each religion. Such angels are given specific names or titles, such as "Gabriel" or "Destroying angel." The term "angel" has also been expanded to various notions of spirits or figures found in other religious traditions. The theological study of angels is known as "angelology." Angels who were expelled from Heaven are referred to as *fallen angels*.

In fine art, angels are usually depicted as having the shape of human beings of extraordinary beauty and sometime androgyny; they are often identified with symbols of bird wings, halos, and light (cc Attribution-ShareAlike 3.0 Unported [CC BY-SA 3.0]).

> As we will learn, people who experience near-death experiences, or comatic introductions to the fourth dimension, report seeing and interacting with angels. This was not lost on St. Thomas Aquinas. Aquinas saw a tension between Pythagorean Harmony and Aristotelian Atheism. Pythagoras, for example, believed God tuned the music of the universe. Conversely, if Aristotle believed in god/gods, that writing is nowhere to be found. However, Aristotle and Aquinas had the metaphysics of First Cause incommon. They both held that there was a quantum #1 "Big Bang" Singularity—a first cause. What they couldn't agree about was form. Aristotle believed in nothing he couldn't bite with his teeth. Aquinas cut for some omnipresent slack. There was an

> interrelationship between the third and fourth dimension that must be taken on faith for the universe to make sense.

Most recent observations at Harvard are beginning to deconstruct the question, Choi (2019) reports,

> If a primordial universe existed before our cosmos, our current understanding of physics suggests massive particles would have existed that would have oscillated at some regular frequency, like a clock's pendulum swinging back and forth. The fluctuations of these "primordial standard clocks" would have resulted in tiny irregularities in the density of matter on minuscule scales that would have become the seeds of structures on cosmic scales in our universe after it expanded.

However, another cosmological model suggests that the cosmos was born from a "Big Bounce," expanding outward after an earlier primordial universe collapsed. (Massive Clock-Like Particles Could Reveal What Happened Before the Big Bang)

So, inflation might have been before and after the singularity of the Quantum # 1 Big Bang, now becoming known as the "Big Bounce." We seem to be pulsating to a frequency with reoccurring hyperinflations—Pythagoras would be thrilled; not least would be Dr. Eben Alexander (Alexander, 2012) *Proof of Heaven* and his embrace of "Sacred Acoustics." He already knew about these frequencies—just sayin' . . .

Plato, Socrates, and Dante support Aquinas

From 468 BCE, Socrates, Plato, and Dante in 1321 AD made statements supporting the duality of the photon, i.e., angel existence. In the 20th and 21st Century, the support continues.

According to PURCH (2019),

> Researchers recently conducted experiments to answer a decades-old theoretical physics question about dueling realities. This tricky thought experiment proposed that two individuals observing the same photon could arrive at different conclusions about

> that photon's state—and yet both of their observations would be correct.
>
> For the first time, scientists have replicated conditions described in the thought experiment. Their results, published Feb. 13 (2019) in the preprint journal arXiv, confirmed that even when observers described different states in the same photon, the two conflicting realities could both be true. [The Biggest Unsolved Mysteries in Physics] (1)

Socrates convinced Plato there were two sides of an interesting statement. Plato used the two at once photon principle to metaphorically get his followers out of the cave. Dante took this to the extreme as he dissented into Hell and made it back to Heaven while writing in striking detail. It is reasonable to speculate that Socrates, Plato, and Dante enhanced their thoughts with psychedelics. A quick "jump-start" into the unseen cosmic order.

St. Thomas Aquinas expended the Summa Theologica to validate his position on the capacity of angels. The remarkable similarity between the Aquinas angel and the quantum photon is generally overlooked in mechanistic science. However, there are defined similarities considering light (bright light) is universally considered significant when experiencing phenomena in the fourth dimension.

Angel

Oxford English Dictionary
A ministering spirit or divine messenger; one of an order of spiritual beings superior to man in power and intelligence, who are the attendants and messengers of the Deity; c. a guardian or attendant spirit; d. figurative, a person who resembles an angel in attributes or actions. (Fox & Sheldrake, 1996, inside cover 1).

Photon

A corpuscle or unit particle of light.
From Gale Virtual Reference Library Photon. (2008).
In K. L. Lerner & B. W. Lerner (Eds.),

> This is the basic unit, particle, or carrier of light. Although it may seem odd that light is made of particles or pieces, it is a fact that light and all other forms of electromagnetic radiation really do behave this way under certain conditions. At the same time, light and other electromagnetic radiation behave exactly like a wave, not a particle, under certain other conditions. Neither description can be abandoned: both are required. Such an object—both particle and wave—is strictly impossible to visualize, but twentieth-century physics proved that light does have this dual or double nature, dubbed "wave-particle duality." However, as we now know, in 2022023 AD its duality can be visualized through experiment.

Architecture reinforces this in tradition:

> Angel light . . .
> Small, roughly triangular light between subordinate arched window-tracery, especially in Perp. Work, are angel lights.

Regardless of where a person looks, the answer always comes in the form of a proton.

> "Heiligenschein, which is German for "holy light," appears as a halo around your shadow as you stand with your back to the sun over moist grass. And the light that creates heiligenschein isn't restricted to reflecting off the grass. You can also spot an airplane's heiligenschein from an airplane window, just around the plane's shadow as it passes over bare fields, as long as the sun is behind it. On the moon, the light that creates it bounces off moon dust—just look at Neil Armstrong's shadow (Klein, 2016).

And, of course, not everyone agrees. Seagrave (2009) responds with "While St. Thomas's *Summa Theologica* remains the locus classicus for natural law theory, this theory acquires more force and persuasiveness when it is placed at the summit of a preceding tradition of thought than when it is relegated to the foothills of an emerging one (na).

In the heat of the argument, movement becomes a question.

In the *Summa Theologica* STI, q. 53, a.1-3, as recorded by Fox and Sheldrake (1996), Aquinas makes his case for the angelic/photon movement.

> An angel is in contact with a given place simply and solely through his power there. Hence [his] movement from place to place can be nothing but a succession of distinct power contacts . . . When [he] continuous it necessarily entails passing through an intermediate place . . . This kind of movement from one extreme of a given place to another, immediately is possible for an angel. Aquinas, in his case for movement, is explaining the photon movement in quantum mechanics. The same phenomena that find itself dancing on the microtubules of Stewart Hameroff and Sir Roger Penrose as they study correlations in consciousness in Orch OR theory. The same revelation is shown to Deepak Chopra in his quantum soup. The same superiorities of morphogenetic fields for Rupert Sheldrake. The same reported phenomena are shown to one in seventeen patients who come and go from general anesthesia. The same light that all near-death experience (NDE) people report with their involvement in the fourth dimension. The same Heiligenschein experienced through visitations into alpha elements and then into dream thoughts, or when standing on the moon. At some point, in all of creation, subjective evidence transcends the deep state academic intelligencia of the hard science academy.

For the Common Man

Members of the scientific community play their information close to their vests, and for obvious reasons. The rest of us are not smart enough to understand their findings or disciplined enough to tolerate their process. There is something about proof that sticks in the craw of hard science. Some people walk in the middle. They are hard science experimenters with a lean into soft science. Here are the explanations of each:

According to Hard and Soft Science (2019),

> Hard science and soft science are colloquial terms used to compare scientific fields based on perceived methodological rigor, exactitude, and objectivity. Roughly speaking, the natural sciences (e.g., biology, chemistry, physics) are considered "hard," whereas the social sciences (e.g., economics, psychology, sociology) are usually described as "soft." The hard science metaphor has been criticized for unduly stigmatizing soft sciences, creating an imbalance in the public perception, funding, and recognition of different fields.

I dislike the iniquitous stigmatization of soft science by the hard science crowd. For reasons yet to be noted, experiments are not exact in the four dimensions necessary to claim the answer is, in fact, accurate. Being in time and space make accuracy impossible. If a scientist lives inside a 40 + trillion light year window from planet Earth, an experiment is only as good as the speed of light. As a practical matter, scientists cannot show present in the experimental process. There is no present. There is only past and future. Everything seen is in the past by the speed of light. The time it takes the photon to reach the observer's eyes, and thereby brain. There is no present.

Maybe a nitpicky point? Well, tend to your knitting, hard science. I like the subjective spontaneity of conversation in the recent past. The would-be soft science.

How Discovery Works

Sir Ken Robinson tells this story.

> I heard a great story recently—I love telling it—of a little girl who was in a drawing lesson. She was six, and she was at the back, drawing, and the teacher said this girl hardly ever paid attention, and in this drawing lesson, she did. The teacher was fascinated. She went over to her, and she said, "What are you drawing?" And the girl said, "I'm drawing a picture of God." And the teacher said, "But nobody knows what God looks like." And the girl said, "They will in a minute."

I fear the ***creativity*** of wondering outside Science, Technology, Engineering and Math (STEM) ***is lost*** on 21st Century members of the Academy.

Astrophysics for the Other

Among today's western icons of the De Rerum Natura tradition is Neil deGrasse Tyson. He is America's unequaled astrophysics celebrity. Tyson brings us the universe through deep education, clever metaphor, and wit. He is the astrophysicist-in-residence with the American Museum of Natural History in New York City, where he serves as the Director of the Hayden Planetarium. He graduated Harvard with a BA in physics, and Columbia with a Ph.D. in astrophysics.

According to Tyson (2017), in his record-breaking book, *Astrophysics for People in a Hurry*, the lesser among us are unable to enjoy the wonders of the universe. That's because we are too busy trying to survive in the day-to-day pressures of life. He follows that with his concern for this uninformed crowd—how his professional life is dedicated to bringing everyone to his threshold of cosmic awareness. So, as he takes the cosmos away, he tries to give it back.

Capacity of Observation

If Tyson pulled his head away from the Hayden Planetarium sphere long enough to see the people, he would be asking a Sheldrake question, "Why is there so much beauty in the Universe?" For Tyson to surmise that the lesser of us are unable to enjoy the universe underwrites the tension between traditional and non-traditional science. Tyson is on the hard. His concern for the collateral damage of war, or the genocide of organized religion, does not reach into the mind of the lesser of humanity. Almost like *throw money at the problem*, Tyson rides on the back of his success to distance himself from the phenomena of the lesser picking flowers and presenting them to local librarians. If there is a demographic in our socially constructed society that enjoys the wonders of the night sky, it's the lesser of us. And, I am not stating the obvious that they are real-time participants in the raw universe. Their routine of wet and cold demonstrates the arrow of time. The homeless know there is a comfortable existence away from where they are; however, to get there requires using a bit of the Penrose Triangle.

Meanwhile, they wonder about the environment. Maybe they don't know the environment stretches trillions of light years, but they do participate in a sea change cacophony of beauty and wonder. They see the adaptation of bees over time being drawn to pollen and nectar in varying degrees between urban neighborhoods. They enjoy the color. They smell the aroma. They become platforms for bees and butterflies between flowers—the lesser of us who know that the sun and the bee and butterfly are somehow related. And, there are 2.4 billion Christians, 1.8 billion people of Islam, and 1.15 billion followers of Hinduism who have the same bee/flower/sun relationship with the universe.

I am using Neil deGrasse Tyson as a straw man in this argument. Tyson is on top of his game. He is as informative and fun to watch, and hear, as anyone in the public space, and honest about the frailty of the human condition. The lesser people—some of them—might benefit from Tyson's efforts at the Hayden Planetarium. Most of them will not.

What I want the Tysons et al. to understand is the lesser people are connected to the universal consciousness of the universe. What all

Sapien science traditions should grasp is that capricious rhetoric toward the lesser for their connection to higher orders is built on feet of clay. Newtonian (hard) science creates information, not joy. To say, "cosmic perspective comes from the frontiers of science" (Tyson, 2017 p. 205) misses the point of being alive.

Perspective is a bigger picture. To be informative in his current format, Tyson might take a break from deconstructionism. Step away from the Periodic Table. Metaphysics gets to work in perception alongside phenomenology. Way more fun and Tyson is about fun as well as science.

The Poetry of Compromise

The string of so many ways to be other than right run throughout this entire text. So, stepping away from the De Rerum Natura approach let's have a look at options. In Buddhism, Dharmacakra (The Dharma) is the path to enlightenment. The Buddha made no claim to the supernatural. His teachings hit the middle ground between east and west. He encouraged people to find nirvana, maybe AKA Heaven. No deconstruction necessary, work both sides of the aisle. Use reason and evidence to enhance thought, but not at the expense of thought. Aristotle thought it was okay to think about thinking. So did Siddhartha—the Buddha.

In the west, the tradition was retelling stories of epic adventures—the tension of good and evil with world order ad hoc to the stories. In the east, there were similar sets of stories. Clans of gods and monsters coming from the Vedas (Lutsyshyna, 2012). Buddha established a path through all the fighting gods with his eight-point plan. If ruminating on the mental compression of hammer-throwing gods, demons, and extraordinary fighting men was an addiction, Buddha was the guy who hit the bellwether for the Alcoholics Anonymous (AA) 12-step plan. The dharma is Right Action, Right Intention, Right Livelihood, Right Effort, Right Concentration, Right Speech, Right Understanding, and Right Mindfulness (Mullin, T. 2012). "Right" in this context is not an exercise in argument and rhetoric. It is a smoothing of the human condition.

The Near-Death Experience and the Lucid Dream

Among the striking examples of Near-Death Experience (NDE), everybody seems to go to Heaven. At least to the light of Heaven. There are some undeniably funny contrarian positions.

Neil deGrasse Tyson speaks of how we should ask questions of people while in Heaven. Don't just stare into the light. Get some Newtonian facts. Where is Heaven? Are you cold? What do you have for breakfast in Heaven? Tyson suggests operating room doctors should post messages on the tops of cabinets in the operating room so when a person leaves their body for the out of body experience; they will read the messages and report the content when they return to the table in their corporal bodies. According to Tyson, the white light NDE people report seeing can be traced to the operating room lights over the table. Tyson makes that point repeatedly. An over-simplification, in my opinion.

There is a famous case in Seattle when a woman left her body while in the ER and, indeed, left the building to see a blue tennis shoe against a third-floor window.

According to YouTube (2011),

> This is the smoking gun of the near-death world! A patient named Maria, at Harborview Hospital (Seattle), had a near-death experience that she told Kim Clark about. She rose up out of her body, outside the hospital, and looked down and saw a tennis shoe on a ledge. Something that could only be seen from a helicopter's view of the hospital.
>
> Kim Clark then went to the floor and looked out the window and couldn't see the tennis shoe. But she reached out the window, and sure enough, there it was, and history was made.
>
> I worked at Harborview Hospital during this period. I personally spoke with the Respiratory Therapist who was involved with the resuscitation. She clearly remembered both the patient talking about the tennis shoe and Kim Clark finding it.
>
> It is SCIENCE that proves near-death experiences are real. Kim's story is supported by science and her own integrity.

Skeptics seem to overlook the zeal with which NDE people report their experience. The International Association for Near-Death Studies Inc. (IANDS) has its oldest chapter in Seattle, WA. In a conversation with one of the IANDS followers, I questioned her story. Maybe her experience was from a hallucinogen. I was quickly informed, she was simply reporting her experience, and my opinions, and or suggestions, were unimportant. This same attitude prevails throughout the NDE population. No amount of science has an impact. This might be why Neil de Grasse Tyson can merely preach to his own choir. NDE people have "seen the light," so to say. And, when one sees the light, nothing so pedestrian as Saipan's questions matter. This sacrosanct position is pervasive throughout IANDS. It becomes the "I know what I saw" position.

Researchers exclaim that nothing is more unreliable than "I know what I saw." Witnesses in court have their stories clouded every day by forensic factor analysis. Phenomena like the Doppler effect can transform the accuracy of witness testimony. On the other hand, courtroom witnesses are usually reporting events they saw under extreme personal stress. The NDE reports are absent the stress. These are happy people on their way to the second most important day of their life—the day they find out why they were born. Recalling "The two most important days in your life are the day you are born, and the day you find out why" (Mark Twain).

Dr. Lloyd Rudy

> This case happened sometime late 1990s early 2000s . . . I did witness the entire case and everything that my partner Dr. Rudy explained . . . I do not have a rational scientific explanation to explain this phenomenon. I do know that this happened. This patient had close to 20 minutes or more of no life, no physiologic life, no heartbeat, no blood pressure, no respiratory function whatsoever and then he came back to life (Daily Grail, 2017).

Dr. Rudy was the team lead for a man having a repair on his heart valve. This required the surgical team to place him on a heart-lung machine. After the surgery, the OR team could not get the patient off the equipment. He would not maintain blood pressure and his oxygen saturation and heart rate would make precipitous drops each time they tried to remove the apparatus.

Eventually, a recovery back to normal pressures seemed hopeless. So, the decision was made to remove the equipment and let the patient go. And he did. All the monitors flatlined. The man was dead.

The operating room filled with staff cleaning up and prepping for another surgery patient. The assistant surgeon closed the body in a manner that would easily accommodate an autopsy—that being a matter of law. People who were hungry went to the hospital cafeteria for some food.

About a half hour passed with the patient having died on the table—the equipment still connected and quiet. Dr. Rudy and his assistant surgeon were standing in the doorway to the OR, arms folded and talking. All the surgical displays and monitors were blank and had been for 30 +/- minutes.

Then, the heart monitor gave a beep. Then another. Then another. Everyone within earshot heard Dr. Rudy scream. GET BACK HERE—GET THE NURSES. WHERE IS THE ANESTHESIOLOGIST!?

All pressures came back to normal with no life support. The patient recovered and was admitted to Post Op.

The medical team visited the patient for about 10 days in Post Op after the operation. The patient had recall of the OR and staff, including Dr. Rudy and the assistant surgeon, for the time he was showing no activity—he was dead. He was apparently in an out-of-body state because his references were from above. He was able to describe circumstances that were available to him only after he was dead. There was no carry-forward information that could be contrived or confused. The patient was out of his body, above the OR staff, observing himself on the table; the procedure to close him for autopsy; Dr. Rudy retrieving Post-It notes on one of the monitors, placed by staff when the patient was in anesthesia.

When questioned, Dr. Rudy became emotional. "All I know is there is some Yang up there pulling strings."

> Some branches of the Yang clan . . . refer to themselves as "Yang of the Hall of Four Wisdoms." The "Hall of Four Wisdoms" refers to a story concerning Yang Zhen, an official of the Eastern Han Dynasty (206 BC–220 AD) and known for his erudition as well as moral character. When a man named Wang Mi visited Yang Zhen at night and attempted to bribe him with 10 catties of gold, Yang rejected the gift. Wang Mi persevered, saying that nobody would know. Yang Zhen famously retorted "Heaven knows, Earth knows, you know, and I know. How can you say

that nobody would know?" Descendants of Yang Zhen adopted the "four wisdoms," or "Si Zhi" as the title of their clan hall. Some Yang family clan halls in various parts of China still carry this name (Yang (surname)).

The Compelling Nature of Personal Experience

Encounters like those of Dr. Lloyd Rudy are refreshing to read. There is a dumbfounding trait among nay-sayers like Lawrence Krauss, Michael Shermer, Rebecca Goldstein, and Steven Pinker that theses sketchy random examples do not meet the test of science. That there was no control protocol in place to verify Dr. Rudy's claim. That the experience cannot be replicated within a control group. And, because of this, the claim is false, and there is an explanation nested in the 200 plus trillion cells making up his patient's body and brain.

In order to arrest an unscientific emotional argument, the God part of anecdotal explanations is countered with the "God is not Great" arguments made famous by Christopher Hitchens. For example, how could a loving God allow thousands of children to become victims of genocide? How could a loving God let my child die too young?

I remember sitting in St. James Cathedral in Seattle when my son Jeff was fighting for his life at Harborview Hospital. I promised God that if Jeff recovered, I would become a flawless Christian. I rocked back and forth while staring at the floor, asking God to spare my son. Did that attempt show my willingness to enter the illusion of the human condition? Hitchens would have reminded me that God was busy, assuming he or she was there at all. Dawkins would have informed me that God is a myth and I would do better seeking some scientific insights toward a medical solution. Michael Shermer might suggest that if I abandon God, my life going forward would be easier because Jeff died through a process of random selection. Nobody is at fault, so have your sorrow and move on.

In retrospect, I admire the Hitchens candor. You will recall from earlier in this text before Hitchens died, his Rabbi asked him if he was going to accept God before he died? Hitchens reply: "Well, I guess this is no time to piss anybody off."

The prospect of a person sitting down to have a conversation with God is remote indeed. Not that it couldn't happen, but what are the odds? Michael Shermer is closer to the answer. People come and go in the organics of the interstellar. Hitchens is right, trying to manage 200 trillion stars and associated planets is time-consuming for God.

Dawkins is right; the time would have been better spent focused on a corporal solution. However, do these suggestions negate the metaphor of God as the higher order? No, they don't. The universe is big, and it's organic. "Every man likes the smell of his own dung" (Erasmus). "Come now, if of herself she is not mad enough, Encourage her" (Terence). "For solid common sense is rare in that estate" (Juvenal). "Just like an ape, man's mimic, whom in jest—A prankish boy in silken cloths has dressed—And left his buttocks and his backside bare—To give the guests a laugh" (Claudian).

I sat daily at Jeff's grave for three years, wondering how this could be? Where he was. If he was alright. There was no answer. Thirty days after Jeff's brother (John) died, I hopped an Air France flight to Paris—walked into Shakespeare and Company—Kilometer Zero, Paris—and bought the complete works of Michel de Montaigne. I read in the shadow of Charlemagne in the courtyard of Notre Dame. And, finally, I realized, as did Lucretius, "For what we have at hand seems best of all." Once I accepted that, I could look away without conditions. Some people call it "flow."

Not long after returning from France, I had a dream. The scene was in sepia tone. There were iconic columns flanking a set of stairs. They were wide and large—like the Spanish Steps in Rome. I was in the dream standing back away. In front of me were my sons Jeff and John. They were running, then somersaulting. They were completely aware of each other and projected happiness. As they ran up the stairs, I tried to get their attention. I couldn't. They kept going up and up the steps between the columns. Finally, they were arm-in-arm and disappeared across the threshold atop the stairs. They were gleeful in their actions. I do not typically remember dreams. I remember that one.

I've thought repeatedly how I felt after the dream. I was relieved. But I was disappointed. I wanted to get their attention. That miss left me wondering if I was as good as I could have been as their father. Why didn't they turn around? Maybe God showed me some mercy sprinkled with a little justice. I don't know, but how I miss them so.

The Capstone Interviews

While researching this book, I found and spoke with three people who represent the overall spectrum of spirituality in the Fourth Dimension. Each one is a stand-alone story of what they experienced. So, I write about them separately. The subjects are Don Martin, Libby Kelleher-Carr, and Dr. Eben Alexander. I spoke with each person, either in person, over the phone, or both. I have an existing relationship with Martin and Kelleher-Carr. I spoke with Alexander for the first time researching for the book. In Alexander's case, however, there was ample Creative Commons material existing, so the Creative Commons Legal Code CC0 1.0 Universal helped a great deal.

In selecting these subjects, I was after two constraints. Each subject must be personally credible. Each story must be extraordinary. In Martin's case, I had a 36-year history with a great deal of day-to-day business involvement. I also know his family. And, as I mention in his copy, his story appeared indirectly. I had no prior knowledge of what he was going to tell me. Because of my maritime background, his story was a windfall surprise.

In Kelleher-Carr case, the same longtime association was in play. I was aware of her story; however, never in detail. Her husband was my corporate attorney, and we traveled together; we found our way around court together. Kelleher-Carr remains the person in her story. For almost 30 years she has been working in financial services and has her own independent life insurance agency. She serves on the boards of NICA (Northwest Intention Communities Assn) and on the Northshore Performing Arts Foundation board. Still an activist, still politically energetic, a mother and grandmother—solid.

Dr. Eben Alexander was a flabbergast. Libby Kelleher-Carr contacted me about meeting him at the International Association of Near-Death Studies (IANDS) Conference. She sent me *Proof of Heaven*, one of his books. In my mind, I dismissed her idea. On her prompt, and a whim, I sent some dailies to his office and forgot about it. In a week, I received an email from *Elizabeth Hare, Assistant to Dr. Eben Alexander*. He was interested and agreed to talk. The following week, at 6:00 AM Pacific Time, we were on the phone together. The man's grasp of our conversation subject was staggering.

Of the three, Alexander has taken a great deal of criticism in the national media. Nay-sayers were, and are, trying to eviscerate his NDE experience. However, in my case, I was fascinated by his descriptive vernacular while in a coma. I felt his words were coming from somewhere else in the arrow of time. And, I think I found out where.

The contrast between the three was that Martin had a lucid dream, not an NDE, but the clarity was mind-blowing. Kelleher-Carr was an NDE that happened over a short period and lingered for 47-years. Alexander was an extended occurrence NDE that is still young in relative terms. I found the mix exceptionally poignant.

Don Martin

During the development of this book, I sent early pages of the text to Don Martin, a writer I know in Arizona. Don is writing after having successful careers upon and across oceans. He, is a graduate of California Maritime Academy and Puget Sound School of Law. He served as General Counsel for multiple large U.S.-flagged shipping companies as well as Director of U.S. Shipping for ConocoPhillips. I may not want to say this in public, but God works in strange ways. He read what I sent him during a business-trip airline flight to Washington, D.C. When he returned, he contacted me by text message writing it would be better if we spoke over the phone. "Holy shit," I thought, I must have breached an important protocol, and the whole project is in the tank!

In a few days, we opened the conversation. I still wasn't clear what I was about to try and defend. Don began with the high elevation of his house, the weather in Arizona, his recent appearance on a radio show in Phoenix—an interview about one of his many articles, and then the conversation stopped—full stop. Finally, I hear, "I have a dream I want you to consider for your book."

I do a prolonged exhale and say, "sure."

"My brother Patrick was killed in an accident in October 1976. We were extremely close; the best I can explain might be the relationship between twins. We were born 13 months apart and for years slept in the same room. Our Mom would often dress not as identical twins as was common in the sixties, but we would wear the same shirts but different colors. Same pants and shoes. Relatives would often tell us that they could not tell us apart. As children, it was not uncommon that we would have the same dream.

"The afternoon Patrick was killed I was out on Lake Washington at rowing practice. I was a sophomore, a history major, member of a fraternity, and a member of the crew team—a typical young man just going to college not really knowing what direction I was heading as far as a career goes but doing what I could to figure stuff out while enjoying college life. I got off the water, and the coach told me to call my father at home. I got the terrible news and drove straight to Gig Harbor, my home. I was devastated, still, am.

"That night I had a dream—from what I understand a lucid dream. In it, I was on the bridge of a white ship with green decks. The ship was in a storm and Patrick was with me. Patrick told me not to worry and that everything would be ok.

"So, two years go by, and I still am looking around for a career as I know, as much as I loved history, it would never make a career for me. A friend told me about the merchant marine (civilian cargo ships) and that there was a merchant marine academy in California. A young man could go there and in four years graduate with a Third Mate Unlimited license and a commission in the USNR. A career in the merchant marine was exciting as I would see the world and such a career would take me through retirement as I would eventually make Captain. I applied and was accepted. I was told to report to campus in August 1978 which would be my first visit to the campus. I arrived on a hot August afternoon and was given a quick tour of the campus and then taken down to the training ship for a tour as well. I had never even been on a ship and was excited. We went aboard and, like most ship tours, went to the bridge first. I was shocked as I looked down on green decks on a ship painted all white. The bridge itself was identical to the one in my dream." (Don Martin, personal communication, April 1, 2019).

So, Don and I finish the call after some normal small-talk, and I start hunting around for comps. I find Patrick McNamara, Ph.D., in Behavioral Neuroscience (Human Neuropsychology) from Boston University. McNamara (2019), calling for the question in Scientific American, "Can Two People Have the Same Dream?" McNamara reports "If two people can share the same dream, then dreams transcend individual minds. There are some commonalities among the reports that increase confidence in their reliability.

For example, most often the two people involved know each other and are emotionally close" (Dream Catcher). Then I went to ProQuest academic library to find almost a million citations under "Can Two People Have the Same Dream? Apparently, many people share dreams with other folks (ProQuest, 2019). And none of this is inconsistent with what Stewart Hameroff and Roger Penrose are dealing with in Orch OR Theory (Volk, n.d.).

We know that much of what we do is computational in cognitive terms. What we don't seem to know, and what Orch OR theory may provide, is how we can have wild and unbridled sex tonight between 9:00 P.M. and midnight, and then consciously relive the entire experience tomorrow at 10:00 A.M. while alone staring into a cup of ho-hum office coffee.

There is also the question of how Patrick Martin got Don Martin two years into the future less than 12 hours after Patrick died. And, how Patrick used the California Maritime Academy training ship "Golden Bear," as his platform, considering Don Martin was a U.W. history major contemplating a career in academics—teaching.

And, here is what makes Don Martin's story solid. If he was to attend the Seattle Physics of Life Foundation Conference on Creative Consciousness, present his story in open forum, and have Lawrence Krauss rise from the crowd to say "everything you say is unscientific claptrap," Martin would raise his eyes to the upper deck, give his brother a snappy-hand-salute, give Krauss the finger, and retire unphased to the lobby bar, where I would buy him a drink.

I followed up with Don Martin on April 1, 2019. We discussed his dream in context with his day-to-day life the past 45 +/- years. I asked him how he handled it with the rest of his immediate family. "I shared it with friends and my other brothers. I never said anything to my parents until much later."

And so, "what was their reaction?" I ask. "I'm not sure how people reacted. I never asked what they thought. I just shared the story." I have got to believe this matter-of-factness is how he stays on top of his game as a captain of industry. Maybe he is the reincarnation of Jack Webb ;-)

I moved the conversation toward his evolving understanding of what happened.

"The older I get, the more I understand what Patrick was saying—'Don't worry—everything will be all right.' His words certainly helped me at that point . . . right after his death, but I think now, some 43 years later, he was speaking to my entire life because things now, at 61 years old, are ok, it's been really tough a few times, but somehow, we all get through it."

I found his evolved understanding insightful considering the time-travel in the dream. Patrick was operating just as St. Thomas Aquinas imagined angels operated in *Summa Theologica*. Patrick was everywhere he needed to be without the interference of time. I asked Don if he remembered bright light on the bridge of the Golden Bear. First, he said "No." Then he reconsidered. "Patrick was wearing a white T-shirt. I remember the T-shirt was very bright—like, too bright to be a T-shirt."

I asked him if the bridge deck was also brighter than one would expect. "No," he said, just Patrick's T-shirt.

All in all, has Patrick's visit with you on the bridge deck of the Cal-Maritime Training Ship been a curse or a blessing?

"Most definitely a blessing."

Then I moved again. This time to how the dream might have affected Don's world view.

"We have absolutely no idea how big our world really is or what's in it. What we see and experience is so two dimensional. I think sometimes there are dimensional cracks and dreams that predict the future."

So, here we were again. Right back with Thomas Aquinas, in the *Summa Theologica*, and his time-traveling angels. Patrick met the Aquinas/Summa test for being an angel. And, it was quick, just a few hours. No extra time in Seraphim school for apprentice angels. Maybe like his older brother, Don, Patrick got right to the point.

So, I finally had to call for the question, "Is there life after death?"

"I firmly believe in existence after death and that we get to see our loved ones again. It's not over after they leave us. Just got to be patient. Easy to say."

I leaned back in my desk chair and thought, "this is the kind of guy you would follow into battle—there is no bullshit—none!"

Libby Kelleher Carr's NDE Narrative

The setting was a place called Hells Canyon, which is part of the Snake River. It forms the border between Oregon, Washington, and Idaho. The year was 1972. Due to his prior commitment, I was there standing in for my boss who was the Northwest Representative of the Sierra Club. Conservationists had already been trying to save this special place for the last 20 years, and here we were, once again, with a bill before the House Public Works Committee. Three congressmen on that committee decided they should come out and see the canyon before making a final decision, one way or the other.

Conservationists saw this as a critical battle since it was the last free-flowing section of the Snake River. What had started off as just a small tour with a few conservationists showing a couple of members of Congress this special place, had suddenly and significantly expanded to include congressional staffers and then the Army Corps of Engineers brass wanting in so they could advocate for building the dam. A professional outfitter had been hired, and now it was a party of at least 30, complete with mountains of duffel bags packed onto three very large Army surplus pontoon rubber rafts, previously used for bridge-building in war-torn areas.

The next day . . .

We arrived on a Friday night and had a get-acquainted gathering before setting upon our adventure the next day. That night I met the ranking member of the committee, James Kee from West Virginia, and known as the "king of strip mining." But somehow, labels didn't seem to matter much at that moment. I remember consciously thinking that I really didn't need to worry about "lobbying" these people because the canyon could speak for itself and the issue was already well-known by everyone present. I just needed to be myself, relax and let the wilderness work its magic.

The next morning, our party started down the river. There were about ten people in each of the three inflatable pontoon rafts—each raft about 25 feet long. Each one carried a mountain of duffel bags. In the middle of each raft, the guide would perch on top so they could guide the raft through the range of slack water all the way up to the rolling boil of Class V rapids.

Not long after we put in, we entered a run of Class V rapids. I had decided I wanted to sit in the bow of the boat so I could get the most exciting view.

Suddenly and without warning, as the raft was violently pitching every which way, I'm looking down into an enormous hole of churning water that just opened up, and the bow just fell straight down into it. This now meant the raft was completely upside down. Almost everyone was thrown out and away from the boat into the middle of the river, but those of us (me and one couple) sitting in the bow of the boat (the fulcrum point) were now pinned underneath it. We all had those old enormous life jackets on with the old steel hooks that were hard to get off. This was years before the quick release plastic fasteners were invented. This proved to just about be my "undoing." Soon Pete and Dee Henault were able to get free from the raft. Then, I was alone.

What had already happened was a massive rush of adrenaline filling every inch of my body. I fought for all I was worth, but I just couldn't get out and away from the raft because the buoyancy of my life jacket kept me pinned up underneath it.

After struggling for a long time, I realized I had nothing left; no air, no energy, and now no will—I was completely spent. I started to relax and let go, and I started taking water into my lungs. The panic was even starting to go, and I remember sort of saying good-bye to myself as I realized with sadness that I was only 23. I realized I had always assumed I'd have a much longer life. But it was apparent that wasn't going to happen.

The Conversation...

Then, the most remarkable thing happened. Just as my mind was fading to black, I suddenly was aware of being enveloped in a cocoon of unconditional love. It was this awareness all around me of a love-energy that had my total attention. It had a presence and a purpose and was there to sustain and care for me. Clearly, beyond all doubt, I had never known this kind of absolute love.

It also imparted to me, instantaneously, a kind of all-knowledge about how everything worked. How Life works, how Love transcends everything—how much bigger and beyond my limited human understanding everything was. I was now being given a beginning glimpse of how everything worked in the Universe. It was indescribable, and I have always been at a loss for words to adequately describe this moment when this understanding came to me.

Note: I later came to feel I was in an energy field that was and is the Source of everything. There are no words for what I was encountering, but it was there, and to this day, I have never forgotten it. To this day, this memory has not faded. And this, in and of itself, is unusual because I often don't remember stuff that happened last week, let alone what happened 47 years ago. This experience was so impactful on all parts of my being that I will remember it until I get to experience it again. But next time, I won't need (or want) to come back.

I then got a question: "Libby, are you ready?" And with that question, it was as if God rested his hands on my head and gave me an even deeper understanding of everything that was going on in the Universe and how it all worked, all at once.

But now with this question, I got the message that this time, it was my choice. I began to ponder the choice I was being given . . . to go or to stay. I remember being very tempted to go, but then I would remember my life back on Earth and that I wasn't sure I was ready to give it up just yet. Then, as I was going back and forth trying to make my decision, another clear message came: "Libby, what is your answer, we don't have all day here!"

With that, I felt I had made a choice, and I started to say "Ok, I'll go."

But, to my own surprise, the "official answer" came out: "No, I can't go, I haven't finished yet."

And with that answer, I was literally given both the Strength and the Will to try again since I had totally spent everything I had as I fought for my life.

With this new infusion of energy, literally on a molecular level, I then tried to push against the raft and the next thing I knew, I was free and able to finally breathe again.

I remember being quite surprised because that was not the answer "I" had decided to give.

Later, after several years of reliving this exchange, I think it was my daughter Molly who "officially" answered that question. She needed me to be her mother this time around (and I needed her as my daughter) which is what happened eight years later.

Finding my friends . . .

Finally, I was away from the raft, just sucking in air, coughing and floating in much slower water. I got myself over to the side of the canyon and just "beached" myself, resting, breathing and surprised that I was unbelievably still alive.

Suddenly, several young Army Corps guys arrive. They had been sent out to look for my body since I was missing from the crowd which had already been gathering down river. They helped me over the rocks to find the rest of our group.

As we came upon the others, I became aware of the congressman who was already grieving the loss of me. I think I reminded him of his granddaughter, but when he saw me, it seemed as if he literally jumped for joy.

And to me, that's just a fantastic part of this story. That people who would typically be political enemies could just drop their political labels. All of that didn't matter. I've never felt so fondly greeted in my life—that somebody was that happy to see me. My cunning political enemy was this gentleman who openly cared about me—the Honorable Jay Kee from West Virginia.

I remember all of us being so happy to see each other that we celebrated and enjoyed the next few days together in that magnificent canyon who had indeed had spoken for herself.

I don't remember anyone discussing what was going to happen to the bill before their committee. All I know is that they went home to Washington, D.C., and passed it. Hells Canyon then gained wilderness status as it became protected under the National Wild and Scenic Rivers Act of 1968.

The aftermath...

For the next few years, I was quiet about what happened to me on the river. I remember only telling my parents about the "conversation" I had had underneath the raft. I remember they listened with rapt attention, and I could tell they believed me, but they cautioned me to keep it to myself and not tell others because it did seem a bit "unusual."

Again, this was in 1972, and nothing had been written about people having these kinds of experiences until Dr. Raymond Moody's book *Life After Life* was published in 1975.

Several years prior to this experience, I took some time out between my freshman and sophomore years of college and ended up living and working just outside of Cambridge, MA at the country's oldest private psychiatric institution—McLean Hospital in Belmont, Mass. While I was working on the admissions ward, I took a course in psychology through the Harvard extension program. As a result of being thoroughly steeped in the psychiatric practices of the day, I knew quite well what happened to people who "heard voices."

As a result, after my experience on the river, I was indeed in a conundrum. Almost every day for three years, I would go for a walk by myself and ask myself "What in the hell happened??" I would argue both sides of the question where I could explain it away by thinking I probably had had a little "psychotic break" because, after all, I was under a great deal of pressure and sometimes people just crack under severe-enough circumstances.

But I never believed that explanation. I would get home from my walk and this familiar debate within myself and just say, "Ok, just tell the truth—what happened??" My answer always and consistently, every day, came back with—"No, I heard the voice of God." And that always felt like the truth.

I finally stopped asking myself that question in 1975 when I read Moody's book *Life After Life*. Raymond Moody was the first psychiatrist to write a book chronicling his research of about 100 near-death experiences. I read the book and often wept because I suddenly was not alone and that others had had similar experiences. And there was now a name for it—the Near-Death Experience. Moody's landmark book describes 10-12 common characteristics of the experience, and I remembered having several of them in my NDE. A lot of characteristics of near-death experiences will overlap even though they will not be 100 percent identical, but they'll have several major elements in common.

I remember feeling an enormous sense of relief to read about his research of other people having had similar experiences, but one of the main elements was they were pretty darn near dying and then there was an NDE. They didn't permanently cross over, but they looked over the fence long enough to have a nice, long look and visit. Reading this book, and later hearing hundreds of other NDE accounts, settled forever my own internal debate about whether I had had a psychotic break. I had not.

Other ways this has influenced my life and world view . . .

I think several things happened to me over time. I think I developed a more vibrant spiritual life that is not focused on denomination or brand of religion. For me, it's focused on my relationship and communication with my Source, that God-Energy I was in the presence of underneath the raft.

Sharing my experience: I find I will openly share my experience when I think I'm in a situation where it may be a benefit to someone else. Perhaps they are facing death or the death of a loved one. I find I'm not at all afraid to go there and engage in those discussions.

However, in the last ten years or so, I have concluded that when and if I choose to share my personal NDE experience, it is not my business as to whether or not I am believed. I don't really care if someone believes me or not because what you believe is your business, not mine. I have simply shared an experience of mine and what it means for my life. If it is interesting or helpful to you, that's fine, but we all get to do our own work.

My advice to others who may have had an NDE: Allow yourself to question it and to process it within yourself which will take years as you relive it and come to understand it more deeply. Also, be open-minded that there was something that might have happened to you that's bigger than your current understanding. And get yourself to an IANDS group to listen, learn and share your story. Or go to websites like www.nderf.org which has collected NDE experiences from all over the world. Do some reading and research and come to know that you are not alone.

(IANDS) The International Association for Near-Death Studies was founded in 1980 by Kenneth Ring, Ph.D. and others who appreciated the impact of Moody's original work on the near-death experience. There are now IANDS chapters around the US and in other countries. (See www.iands.org). This is only the beginning (hopefully) of people wanting to have a place to share their experiences and to better understand them. It is reliably estimated that about 17 percent of the American population has had an NDE. Many others have had lucid dreams which share similar characteristics to the NDE, but are technically different in that the participant is not actually quite near death, or crossed over and not too much later, has returned.

So, after about 29 years of attending these meetings, on and off, I have found my tribe. I find a great deal of comfort in hearing the NDE stories of other people. And after discussing the similarities and differences, it's amazing how many of us have arrived at many of the same conclusions about Life.

After many years of examining my experience, I came to believe in reincarnation as the model of a concept, a metaphor, that we go through in search of a way to understand our human experience and the world. I came to think of it as being in school, and that we have to keep repeating our experiential lives until we get it pretty nearly "right." We get to matriculate when we have learned the major lessons which are to Love and to Grow, and also probably to Forgive before we can be released and go onto the next plane of existence which I believe will be wonderful.

An afterthought: Lucid Dreams

What else happens is you recognize the essence of energy. At least 10 years after my NDE, I gave my mother "permission" to leave this life and go to the other side. She had had lung cancer and was very weakened by the radiation treatments. The day after Thanksgiving 1985, she fell and gave herself a compound fracture at the ankle. The ambulance came and took her to the same hospital she had worked at for 24 years as a medical technologist in Eugene. I came down from Seattle the next day. I walked into her room, and I finally got a chance to ask her if she understood her condition and that the doctors didn't think they could do surgery because she wouldn't survive the anesthesia.

And she said, "Yes, I understand what's going on with my leg." I then responded, "Mom, what do you think should happen? What do you want?" No one had asked her that question. Everybody was a cheering squad because they didn't want to lose her. It hadn't been about what she wanted.

She simply said, "You know, I think it would be best if I just left." I said, "Well, Mom, that's okay with me. Dad and Rayl will be okay. We'll all be okay if that's what you want to do; that's okay." And then she said, "Will you tell me what happened to you again that day underneath the raft?" So, I told her my story again, and she appreciated it; she was ready, and she died that night. I was with her just before she passed. It was just fine.

Two weeks later, she appeared next to my bed on the "Sundown," which was our boat we lived on in Seattle. She was standing right next to my bed in the early hours of the morning when I was first waking up. She looked amazing because she was radiating light as she gave off waves of happiness. She was radiating life and light, and she looked like she was 23 years old. Then she said, "I just wanted you to know I'm doing fine, having a great time, wish you were here." "Oh, okay. Well, I'll see you soon enough." Like she was on vacation in Hawaii or something. And then she communicated to me that she wanted me to tell my brother and my dad that I had had this conversation with her and that she was doing just fine.

I honored that request and told them both about my visit with her and what she said. What was really nice was that they both believed me. They didn't question what I said nor my sanity. They didn't question anything. They just received the message and were happy to receive her communication.

I understand what I had with my mom was a lucid dream experience. I've seen this referenced in the NDE literature, which is described as an experience of a very real, almost hyper-real intense dream, often coming with a specific message from and/or to a loved one.

Note: I want to thank John LaCasse for inviting me to tell my NDE story which he plans on using in a book he is writing about consciousness.

I am very grateful for this invitation which has made me focus on what is probably my most significant life event. Having to retell it has made me reach a renewed appreciation of it.

It always felt to me like I got to cheat on the final exam and was given the answers first to life's big questions in advance. Not that I always understood them, but at least I got to peek, early.

LKC

Dr. Eben Alexander

In August of 2018, my friend Libby Kelleher-Carr called me to say she attended the International Association for Near-Death Studies (IANDS) National Conference in Bellevue, Washington. An organization specific to interest in Near-Death Experience (NDE). She told me, she met and had a conversation with Dr. Eben Alexander. I was deep in projects with Silvertip Research and had no idea who Dr. Alexander was. That seems remarkably narrow as I type these words, considering his universal profile, his books, and not least, his crossing into the afterlife. One would think someone with my curiosity would know about this guy.

Libby told me I had to learn about Dr. Alexander and that she was sending me his book, *Proof of Heaven*. I agreed, and as a preliminary to receiving his book, I went to YouTube and found him giving speeches on a global scale. The book arrived by USPS with a blue butterfly embossed on the cover, and seven high-end endorsements inside Cover 1, suggesting I was about to change my life.

His writing style reminded me of Lisa Genova and her treatment of *Still Alice*. Poignant, and once in hand is not put down. I'm not qualified to discuss his near-death experience brought on by a rare form of meningitis. The opening of the book is chock-a-block with medical jargon and the acronyms that go with them. However, a non-medical person comes away with the reality that this fellow was in deep trouble—indeed, on his way out.

As I reflect on NDEs today, I find both Dr. Eben Alexander and Don Martin have a hook in their experience ratifying their stories in practical terms. I think that practicality is what people want to find when reading about stuff on the "other side," i.e., the fourth dimension. They look for some circumstance that does not appear contrived and the person's resolve toward the experience. And, Libby Kelleher-Carr's experience illustrates today's unyielding IANDS person, who says: "I am simply telling you my experience." Then they walk back toward the light—utterly fearless!

What made Dr. Alexander unique was his background as an academic and practicing physician. He was trained to think in the frame of the scientific method. In Martin's case, he had a great deal of life experience in maritime, business and legal as a solid citizen, and he spent five decades underplaying his story. It's like his story was parked in his garage for 50 years under a canvas tarp. These circumstances added to their credibility.

Here is a truncated segment of Dr. Eben Alexander speaking at Westminster Town Hall, Minneapolis, Minnesota U.S.A.

> *This is how it began. I woke up at 4:30 in the morning November 10, 2008, with severe back pain . . . it was soon thereafter that I slipped into coma . . . and was rushed to the emergency room where an astute physician realized I might have meningitis.*
>
> *The ER team did a lumbar puncture and [Delete: when] the fluid surrounding brain and spinal cord came out of my back as pus, under pressure . . . my doctor later told me that when she saw that, she knew I was dead.*
>
> *In coma . . . a . . . perfect pure white light with fine white and gold filaments that open like a rip in the fabric around me. Into a lovely gateway valley. A perfect ultra-real valley lush with life. And I was moving up through it because I was a speck of awareness on a butterfly wing with millions of other butterflies.*
>
> *Lots of souls dancing below us—joy and mirth. Beside me—on the butterfly wing—a beautiful girl . . . And she never said a word this beautiful girl on the butterfly wing with sparkling blue eyes and high cheekbones, high forehead, wide smile. But her thoughts came straight into my awareness, and these I share with all of you because they are important for all of us.*

Normally, these descriptors are not common in our day-to-day activity. We would not describe ourselves as [a] "speck of awareness . . . among millions of butterflies." So, Dr. Alexander is describing a state of wonder and amazement that is trenchant to his being. He is laddering somewhere in the canopy of higher orders, if not butterflies.

> *The message from her was very clear. Unconditional love, infinite love of the creator. And in fact, in that beautiful realm . . . amplified because the joy and mirth and all these souls . . . fueled by the scooping orbs of pure gold and spiritual beings up above. Leaving sparkling trails and emitting these anthems, crescendo after crescendo of the most beautiful.*
>
> *Of pure conceptual flow and of love and of emotional power and that gateway realm with the unconditional love and the beauty and joy down below me with all those spiritual beings, and the swooping orbs of angelic choirs above . . . the angelic choirs above, offered a traversal into higher and higher spiritual realms . . . ascending and seeing every bit of this realm collapsing down. Even the spiritual realm and what I call deep time a higher causality collapsing down and going out into the core. Infinite inky blackness but filled to overflowing with the divine love of an infinite being of compassion and mercy.*

As I read and listened to Alexander's Westminster Forum speech, I kept feeling I know this from somewhere. Regardless, as I read and listened to Dr. Alexander, I couldn't get his words out of my mind. Like I have, somehow, been here before. I was listening to this physician, just returned from Heaven, who was talking in a pentameter reminiscent of Classical Greek culture. In the live presentation, I could see some takes of the audience. They were visibly dumbstruck. Who is this guy, they were thinking? He was working in "Daktylos" (Dactylic) pentameter like he was standing in a Greek Basilica.

The Greeks equated harmony and calculation. And the body was the foundation of art and science. Look at your finger. There are three joints. From the upper—two long and one short. It was, Higgledy-piggledy, boo., Higgledy-piggledy, boo., and off his tongue are coming these souls, of scooping orbs, of pure gold, of spiritual beings, of sparkling trails and anthems of crescendo after crescendo and >> Higgledy-piggledy, boo . . .!

> The hexameter was first used by early Greek poets of the oral tradition, and the most complete extant examples of their works are the Iliad and the Odyssey, which influenced the authors of all later classical epics that survive today. Early epic poetry was also accompanied by music, and pitch changes associated with the accented Greek must have highlighted the melody, though the exact mechanism is still a topic of discussion.
>
> The Homeric poems arrange words to create an interplay between the metrical ictus—the first syllable of each foot—and the natural, spoken accent of words. If the ictus and accent coincide too frequently the hexameter becomes "sing-songy". (Creative Commons Legal Code CC0 1.0 Universal).

So, I threaded my fingers through the hair behind my head, and I looked outside. There was a crow watching through my open office window. I asked the crow what might be going on here.

"This guy is Pythagoras!" says the crow. "He's in Croton, Italy speaking in the harmony of whole numbers." Then the crow said, "Go find the 17th-century diagram by Robert Fludd of the Pythagorean Universe. You will notice that the hand on the scroll, tuning the universe, is the hand of God. And the whole shebang is in hexameter harmonics of Dactylic pentameter. Higgledy-piggledy, boo . . .!"

Before I could respond, the crow hopped into the air—away for the night. I called out to the crow, "Why are you leaving?" And I take notice in cutting Doppler effect, "Because I figure you still have that damn BB-Gun."

What follows is Pythagoras, in the *Life of Pythagoras* by Iamblichus of Chalcis, describing the first erudition rhythms of the soul:

> He arranged and adjusted what might be called preparation and touchings, divinely contriving mingling of certain diatonic, chromatic and enharmonic melodies, through which he easily

> switched and circulated the passions of the soul . . . Each of these he corrected by the rule of virtue, attempering them through appropriate melodies, as though some salutary medicine (Guthrie, Fideler, & Joscelyn, 1940, p. 72).

Dr. Alexander is describing, as he witnesses it, the Pythagorean harmony of the organic universe. He is figuratively somersaulting through the transcendence of the arrow of time. And, he later admits that all of this is a composite of "top-down" light and sound. That he is in [a] Pythagorean harmonic of all that is created. He manages to say that "for me, the word God was too puny a word." Dr. Alexander is overpowered by the amped-up power of the universe. However, that power is not stuff. It's like I wrote before. Take the Pyramid Amlong Crystal and place it in the sun. What flashes on your ceiling is the spectrum of visible and invisible light that is responsible for everything we recognize—the sun! However, the spectrum does not look like your furniture, your car, or your cat.

Dr. Alexander was witnessing the back story of the same spectrum, including the Pythagorean scale of sound. He was at the source. His arrow of time may have made the transverse back in measure 2500 years.

The Collaboration with the Higher Order

In the opening pages of *Fight for the Quantum*, we discuss a few mathematicians of consequence who lost their bearing when they could not justify the mechanics of their discipline. In the ending pages of this book, we examine people of consequence whose experience transcend common reality placing them at odds with brain-centered consciousness because their brains were dead, and they kept going.

Most recent among them in the high-profile arena is Dr. Eben Alexander. His story is compelling enough that if he had described Heaven, not as love, light, butterflies, and golden orbs, but as sext-trillions of Echo Dots, with Jeff Bezos as God, we would have believed him. Not because his vision was so different, but because his credentials are so deep. The same public situation seems to prevail with military pilots who see flying saucers. If a flyer has a U.S. flag patch on his or her shoulder and sees a strange object operate at right-angle warp-speed in the sky, it's "Game On" the aliens have arrived.

The consistency among the seers of out of the ordinary becomes a testimony to their personal impact. The stories do not change. These people know what they saw, they continue to feel the experience, and, as with the public spectacles in the Roman Colosseum, the thumb of the Emperor can go either way—it makes no difference. It is this Expérience de mort imminente NDE phenomenon that, according to Macisaac, (2014), reports,

> 13 million Americans, or 5 percent of the nation's population, had experienced an NDE as of 1992, according to a 1992 Gallup

> poll cited by the Near-Death Experience Research Foundation. 774 NDEs per day are experienced in the United States, according to the same poll. That means another 6 million or so Americans may have had NDEs since the 1992 poll, raising the number of Americans who have had an NDE from 13 million to 19 million (1).

In perspective, that means the total populations of New York, Los Angeles, Chicago, Houston and Philadelphia (combined) could be NDErs. Obviously, NDE people are not campaigning in the streets. There is no lobby organization working the parties on Washington, D.C.'s Embassy Row. These guys are playing it close to their vest, so to say.

The other side of that coin was in 1977 when George Lucas decided to let the FORCE BE WITH US. There was a $10 billion-dollar preconditioned market. When the phasers snapped-on, NDErs were relieved, not surprised.

The Way Forward

So, let's compare acronyms. The National Rifle Association (NRA) has five million members. They are known around Washington, D.C., as the most effective lobbying organization in the history of the United States.

In contrast, there are about 19 million NDErs in America. That's bigger than the NRA by a factor of four (4). What if NDErs decided to organize? Not sit and listen to companion stories—organize. The American Federation of Labor and Congress of Industrial Organizations (AFL-CIO) organize. The American Federation of Rapture and Cross Over Experience (AFR-COE) organize.

Look at the Difference

The International Association for Near-Death Studies (IANDS) supports the world knowledge from NDE research and scholarship wherever compassionate people educated by that information can offer understanding and support to their community.

The American Federation of Rapture and Cross Over Experience (AFR-COE) is a federation of 70 national and international Rapture Unions that represent 20 million men and women whose main purpose is to bring transcendent formality through social and economic justice.

So, Am I Joking?

Maybe, maybe not. The NDE people are already fearless about their experience. They are simply having a hard time knowing what to do with it. For society to move beyond material science, the hypothetical AFR-COE socially constructed order must accumulate power. That must be power based on their strength, not fear. Today, the mechanism community looks upon NDErs' as did the Roman emperors—with disfavor toward their heresy—willing to vocally strike them down with contention. There is nothing in Saipan's 70,000-year history that suggests NDErs' standing will gain credibility without the use of power-abstractions against mechanistic science.

Where is the Power?

The power rests in defense against statistics-mining. Materialists have mathematics on their side. They demonstrate, through the scientific method, that the Sapien consciousness resides only inside the skull cap. They use data-sets measured in billions demonstrating the actuality of mechanism while knowing that "no neuroscientist on Earth can give the first sentence to explain a mechanism by which the physical brain gives rise to consciousness" (Alexander & Newell, 2017, p. 41). They ignore the 774 plus people per day in the United States who experience the opposite reality, at least according to the Gallup organization. And those 774 people must develop an attitude.

For example, let's say I am carrying a Smith & Wesson +P 38 caliber revolver for protection. A fellow says of my carry-pistol, "That S&W 38 you carry is not much of a fight stopper." I say, "Okay, how about you step over there about 10 feet. I will double-tap (two shots) your ass with my 38. Then you come back over here and tell me how that worked out for ya." This is about positioning attitude. It's not about winning. It's about maintaining one's own course and speed.

"We cannot teach people anything; we can only help them discover it within themselves." —Galileo Galilei

I am curious which iconoclast of empire might have followed that advice. Mother Teresa, Napoleon, George Washington, Mahatma Gandhi,

Nelson Mandela, Martin Luther King Jr.? None of them. They all had a hard-fought position away from ruling party hegemony with a form of disobedience. Not to say the remains of empire are worth anything. One can be effective without a garrison of stuff. I have this vision of a funeral ceremony at my "Take it with you Cemetery." There is an enormous knuckle crane lowering a coffin into the Earth. The hole is yawning like an open pit mine. There are dump trucks—the big ones with huge wheels—circling down into the pit carrying all this guy's stuff. Mourners are on bleachers around the pit. They have binoculars. After the coffin is in position and the trucks are unloaded, dozens of bulldozers come in unison pushing dirt over the stuff and into the hole. Ticketmaster sells tickets. It's my Monster Truck Death Jam.

So, yeah, there is internal discovery. At least some. But there must be a call to the cause. During the Second World War, there were "Code Talkers." Special people who spoke Indigenous Native American language. They were talking past the Nazi listeners with engagement instructions for the allied forces. They were special. Keep in mind they were "talkers" not "listeners." Currently, NDErs tend to be listeners. There are exceptions. Any NDE trip to YouTube will make that clear. However, like many adventures into the unknown, countless people who experience something in the fourth dimension keep it to themselves. Not to be branded as quacks—maybe dangerous. But these are the people who are set apart for some purpose. These are the people who have already round-tripped into the celestial hierarchy. These are the people—the designated ones—the travelers who watch Doctor Who as a travelogue rather than science fiction.

Accordingly, in (Wikipedia, 2019),

> Thirteen actors have headlined the series as the Doctor. The transition from one actor to another is written into the plot of the show with the concept of regeneration into a new incarnation, a plot device in which a Time Lord "transforms" into a new body when the current one is too badly harmed to heal normally. Each actor's portrayal is unique, but all represent stages in the life of the same character. Together, they form a single life with a single narrative. The time-traveling feature of the plot means that different incarnations of the Doctor occasionally meet. The Doctor

is currently portrayed by Jodie Whittaker, who took on the role after Peter Capaldi's exit in the 2017 Christmas special "Twice Upon a Time" (1) Seem familiar? (Creative Commons 2019) https://en.m.wikipedia.org) (Creative Commons Legal Code CC0 1.0 Universal). (cc Attribution-ShareAlike 3.0 Unported [CC BY-SA 3.0]).

Is There a Prosaic View of NDE?

Therein may be the chink in the armor. Human nature prevails, even though these people have already been to Heaven. In each case reliability is important. How important is acceptance among our peers? For the last 17-years "Stolen Valor" has been on the rise. Principally, men who have never been in the military manage to get a uniform, buy campaign ribbons, and troll public spaces for acceptance and admiration. Typically, they are outed by an active or inactive member of the armed forces who spot the incorrect placement of campaign ribbons, along with oddball shoulder patches that don't sync with the uniform. This kind of desperation might happen in the highest orders—the hand of God. And, considering there is limited source material, any story will do. The problem being spotted based on the number of tries for recognition, using the same story among strangers.

Do People Hear Us in Heaven?

Richard Martini writes in QUORA reprinted here under Creative Commons license.

> I'm only weighing in with the research in the topic. This is not my opinion, belief or theory about "how do people who are dead hear us?"
>
> I've taken out the word "Heaven" in this reply, because it has a religious connotation. If someone is asked "What is Heaven?" almost everyone will reply with some version of what they've been told Heaven is.

In the thousands of deep hypnosis cases I've examined (and the 45 I've filmed) people don't refer to "Heaven" as the place we "go" after this lifetime. They consistently use the same word to refer to this place . . . they call it "home."

They don't call it "home away from home" or that place where "only good people go." They consistently say "home" when asked "Where would you like to go now, after your memory of a previous lifetime?" or "I want to go back home" or "I want to return home."

So, if we use an alternate word for Heaven (it's okay to use Heaven, I'm just saying it causes "brain freeze" in many people), "How do people "back home" hear us?"

They say that it's a matter of frequency. Not just the frequency that is related to our own time and space—but their frequency as well. I've interviewed dozens of people about the process—people both here and "off the planet" and what they say is consistent. (If they were making it up, or the various mediums I was working with were making this up, they would have casual mundane answers, or a variety—but they don't. They consistently say, "it's a matter of math" or "it's a matter of frequency.")

Imagine that you're trying to converse with a hummingbird. In your mind's eye, the bird is in, out, gone—in the hummingbird's point of view we are moving incredibly slowly. Or think about it in terms of looking into a fish tank. We can see the fish inside blinking, looking around, but in general, they observe us as light on a glass. We don't think we can "speak" or "communicate" with animals, but we do all the time.

Take, for example, cats who see things that aren't in the room. I've watched two cats in a room where their owner had just crossed over—and for 15 minutes, they watched "something" in unison flying around the room. Or animals that can smell cancer. We can't smell cancer, but they can. We can't hear the high-pitched sounds, but they can. We can't smell a dinner a mile away, but they can. Just because we can't understand the frequency with which bees can see UV light—that doesn't mean they can't see UV light.

So, let's begin there.

Every sensation that comes into our consciousness—light, sound, taste, touch—everything is a frequency or a wave that gets translated into signals that are translated into information that we process. There is no "screen" in front of me as I type this—there are atoms that have come together to hold that space. There are ones and zeroes that I'm using via a keyboard to impart meaning and syntax. These dots or pixels have no inherent meaning—but to a person reading it, they might.

People who can communicate with people no longer on the planet do so because they can get a glimpse of the frequency—they can hear it, see it, sometimes feel it. They don't always get exactly or precisely the information that is being imparted—but sometimes they do. Sometimes they "hear" or see the information they're seeking. I work with a medium who works with the FBI and other law enforcement agencies in missing person cases, and her daily routine is to "see" or hear or experience things that we can't.

She's been accurate enough that the head of NYPD flew her into NY to find out why she knew so much about the death of a prominent citizen as if only she could have witnessed those events. (Bill Bratton) But she does her work pro bono for law enforcement and helps connect people to those who are no longer on the planet.

What I focus on is "new information." So, if I'm "interviewing" someone who is no longer on the planet, I ask questions that I can't know the answer to. "Who was the first person to greet you when you crossed over?" I can then go to the family of that person, ask for the identity of that individual and verify other details about what I've heard in the interview. I've been doing this for three years now, and it's pretty amazing what the new information is.

But I'm answering this question not to upset anyone's apple cart. It's to state simply that I have been researching this very topic for a while now—and based on the "new information" (that later turns out to be accurate, verifiable, that could not be hypoxia, a hallucination or some form of cryptomeria or synesthesia) I track down the details of that report. Sometimes I'll ask the same questions to the same individual (whose life I'm familiar with) with different mediums to see if I get contrary information. I have not.

All I can say is that they consistently report that they are trying to reach out to us. They are having a hard time communicating with us. Because they know that they still exist—and are frustrated by the fact that we believe they do not. They have a hard time reaching out to us because we don't take the time to adjust our frequency to hear what it is they're trying to tell us. This isn't just one account, but dozens of accounts from different people who claim that "over there, they're fine, enjoying the view, but are frustrated by trying to reach out to their loved ones and are unable to do so because their loved ones can't hear them."

It's not a matter of belief. It's a matter of opening ourselves up to the possibility that there might be some realm or form of existence after this one. We see it in caterpillars. We see it in nature often, but for some reason, we are convinced that butterflies don't exist, can't exist—because how could a crawling bug become something else? We tend to focus on the chrysalis—worship the cocoon, build monuments to it—when we choose to ignore the butterfly sitting on our shoulder shouting "hey! It's me! I'm still here! Stop worshipping my cocoon, would ya and turn around!"

This is not my belief, theory or religious philosophy. I'm just reporting what people consistently say—and the results are reproducible (if people take the time to reproduce them).

How do we talk to Heaven? Easy, just begin asking questions. As I'm fond of saying when you go into a church and talk to someone who isn't there, it's normal; if they reply, that's when it gets complicated. The next question is "What do you want to say to your loved ones no longer on the planet?" The answer to "How are you?" is almost always "I'm fine." The answer to "Are you dead?" is "No, I'm not. How could I be talking to you if I was?" The answer to "Where are you?" is often "I'm right here."

So, I recommend thinking a bit about what questions you want to ask to folks who are no longer here. "Who did you see when you crossed over?" You don't know the answer to that question, but they do. "Is there some kind of signal or sign that you can use to reach out to me, a feeling or a sensation I can get to know you're nearby?" "What do you miss about being on the planet?" "What do you regret if anything?" "Who are you hanging out with and what are you doing?" These are all unusual questions, but I ask them to everyone that I interview. The answers are always fascinating! (Creative Commons Legal Code CC0 1.0 Universal).

Sometimes It's Truly Funny

I will leave this writing with a couple of stories of my own (the names and locations are changed for obvious reasons) "Blue Seraphim Comes Back as a Horse" and "God tells The Enlightened One to eat potatoes." There are few things more relaxing than writing about something one enjoys, especially if the topic is wide open with no authority of consequence.

What could be better? God, in his or her wisdom, provided us with the ability to think in abstractions about 70,000 years ago. In language, we developed the Fourth Dimension. That place where we create our interpretation of the universe.

Blue Seraphim Comes Back as a Horse

The New Church School of Illumination is an American spiritual sect near a small Atlantic coast town. The school was established by a person, who claims to channel a 70,000-year-old Blue Seraphim, known as The Illuminated One. The school's teachings are based on channeling the Blue Seraphim. As with many political, social, or spiritual movements, there is a need for money from time-to-time. In this case, I was in my office in New York when this man walks in with a strange request. He was looking for a Mickey Tug. He knew, from somewhere, I held a Merchant Marine Master License and, as a consultant, might be able to help him. The Mickeys were between 39' and 160' built for harbor service in the Army—WW 2, especially. As luck would have it, I knew where to find a big Mickey. I suppose he knew that in advance, otherwise, why me?

Being curious, I wanted to know about this fellow and his use for the tug. He said it was a "special-use purpose" and wanted to know about carpenter/shipwright availability where the tug was berthed. Not a problem, and in a few days, the Mickey was his. As I recall, the feature reporter for the journal of the small town noticed all the activity aboard the tug and ran a story. Big news then because the boat had become a major source of employment in a down economy. The newspaper story was about employment, not the upcoming event in the Mickey's queue.

The new owner was a follower of the New Church School of Illumination. As the story goes, the New Church School proctor told this guy that Blue Seraphim was planning a return. Blue Seraphim planned to come back as a horse, and he wanted to have an Earthly offspring. (Sort of a Jesus Christ knock-off) The copulation should be with an attractive blond woman of some healthy consequence. The New Church School proctor opportunely happened to own the horse that Blue Seraphim identified as his incarnate self—at least long enough to mount an attractive blond of healthy consequence. So, my buyer, seeming to be of some consequence himself, decided to buy the horse and manage the entire event. That included the Mickey tug whereon the shipwrights in Small

Town were fabricating a scaffolding around the fantail of the tug so Blue Seraphim (The Horse) could be positioned to have his way with a blond woman of healthy consequence—yet to be named.

By now the story is circulating around town and around the state. The rumor was that the horse went for a quarter million dollars—the new owner was delighted. I expect so was the New Church School proctor. Easy to understand why. For a mere quarter million plus another hundred thousand for an aging Mickey Tug, he got to be close with the Illuminated Being, Blue Seraphim. And then, of course, there was the scaffolding around the fantail, the cranes for the horse, the viewing stands—oh, did I forget to mention those? And the capstone event.

So, I am back in my office, and these two men walk in. They are dressed in midnight blue Hickey Freeman suits, black Coal Hawn shoes, and John Nordstrom ties. They are lawyers. They are looking for me. We talk. They said they represent a multiple generation Fortune-500 American conglomerate. They understand that I recently sold a large boat—a tug, they understand. I said "yes." "Well," they say, "your customer is in the family of the F-500 Co." A distant relative of old F-500, I presumed. My customer/buyer turned out to be a trust fund junkie who was burning through the Corp Trust money and F-500, et al. were not happy. Don't know what happened after that. Lost track of the boat and the horse. Never met the blonde. I did spend some time with Old No. 7, Jack Daniels Tennessee Whiskey, wondering if I had anything else in inventory the Illuminated Being Blue Seraphim might need.

God Tells the Enlightened One to Eat Potatoes

Another time, this young portly guy walks in with a cadre of followers. He is this group's enlightened one. He wants a boat—a big boat—which he will convert into a church, and float around dispensing the Hand of God. I find a freezer ship, a decommissioned military. We do the deal on a 10-year contract. After we close, I visit the ship. There are trucks on the dock. They are full of potatoes that are being loaded on the ship's weather deck. The new owner is on the dock. I ask about the potatoes. He tells me, "Potatoes are the only food identified by God to be eaten by man." I go back to my office, but, this time, I look around for some Irish Whiskey.

The Players

Eben Alexander

An American neurosurgeon and author, Alexander's book *Proof of Heaven: A Neurosurgeon's Journey into the Afterlife* (2012) describes his 2008 near-death experience and asserts that science can and will determine that the brain does not create consciousness and that consciousness survives bodily death.

Alexander is also the author of the 2014 book *The Map of Heaven*, which builds on the claims in his previous book, and coauthor of the 2017 book *Living in a Mindful Universe*, which describes his personal journey since 2008. Alexander was born in Charlotte, North Carolina. He was adopted by Eben Alexander Jr and his wife Elizabeth West Alexander and raised in Winston-Salem, North Carolina, with three siblings. He attended Phillips Exeter Academy (class of 1972), the University of North Carolina at Chapel Hill (A.B., 1975), and the Duke University School of Medicine (M.D., 1980). He completed his Neurosurgery residency at Duke University Medical Center in 1987 followed by a Cerebrovascular Neurosurgery fellowship at the Newcastle General Hospital in the United Kingdom in 1988 (Creative Commons Legal Code CCO 1.0 Universal)

Neil deGrasse Tyson

Tyson studied at Harvard University, the University of Texas at Austin and Columbia University. From 1991 to 1994 he was a postdoctoral research associate at Princeton University. In 1994, he joined the Hayden Planetarium as a staff scientist and at Princeton, in 1996, he became director of the planetarium and oversaw its $210 million reconstruction project, which was completed in 2000.

From 1995 to 2005, Tyson wrote monthly essays in the "Universe" column for Natural History magazine, some of which were later published in his books *Death by Black Hole* (2007) as a visiting research scientist and lecturer, and *Astrophysics for People in a Hurry* (2017). During the same period, he wrote a monthly column in *StarDate* magazine, answering questions about the universe under the pen name "Merlin." Material from the column appeared in his books Merlin's Tour of the Universe (1998) and *Just Visiting This Planet* (1998).

Tyson served on the 2001 government Moon, Mars and Beyond commission. He was awarded the NASA Distinguished Public Service Medal in the same year. From 2006 to 2011, he hosted the television show NOVA Science Now on PBS. Since 2009, Tyson has hosted the weekly podcast Star Talk. A spin-off, also called Star Talk, began airing on National Geographic in 2015. In 2014, he hosted the television series Cosmos: A Spacetime Odyssey, a successor to Carl Sagan's 1980 series Cosmos: A Personal Voyage.

The U.S. National Academy of Sciences awarded Tyson the Public Welfare Medal in 2015 for his "extraordinary role in exciting the public about the wonders of science." (Creative Commons Legal Code CC0 1.0 Universal).

Sean M. Carroll

He has appeared on the History Channel's The Universe, Science Channel's Through the Wormhole with Morgan Freeman, Closer to Truth (broadcast on PBS), and Comedy Central's The Colbert Report. Carroll is the author of Spacetime and Geometry, a graduate-level textbook in general relativity, and has also recorded lectures for The Great Courses on cosmology, the physics of time, and the Higgs boson. He is also the author of three popular books: one on the arrow of time entitled *From Eternity to Here*, one on the Higgs boson entitled *The Particle at the End of the Universe*, and one on science and philosophy entitled *The Big Picture: On the Origins of Life, Meaning, and the Universe Itself.* He began a podcast in 2018 called Mindscape, in which he interviews other experts and intellectuals on a variety of science-related topics (Creative Commons Legal Code CC0 1.0 Universal).

Brian Cox

Brian Edward Cox OBE, FRS is an English physicist who serves as professor of particle physics in the School of Physics and Astronomy at the University of Manchester. He is best known to the public as the presenter of science programs, especially the Wonders of . . . series and for popular science books, such as *Why Does E=mc²?* and *The Quantum Universe*. He has been the author or co-author of over 950 scientific publications.

Cox has been described as the natural successor for BBC's scientific programming by both David Attenborough and Patrick Moore.

Before his academic career, Cox was a keyboard player for the bands D: Ream and Dare (Creative Commons Legal Code CC0 1.0 Universal).

Richard Dawkins

Dawkins first came to prominence with his 1976 book *The Selfish Gene*, which popularized the gene-centered view of evolution and introduced the term meme. With his book *The Extended Phenotype* (1982), he introduced into evolutionary biology the influential concept that the phenotypic effects of a gene are not necessarily limited to an organism's body but can stretch far into the environment. In 2006, he founded the Richard Dawkins Foundation for Reason and Science.

Dawkins is known as an outspoken atheist. In interviews, he has called himself an agnostic about many matters of religious faith, instead endorsing reason. He is well-known for his criticism of creationism and intelligent design. In *The Blind Watchmaker* (1986), he argues against the watchmaker analogy, an argument for the existence of a supernatural creator based upon the complexity of living organisms. Instead, he describes evolutionary processes as analogous to a blind watchmaker, in that reproduction, mutation, and selection are unguided by any designer.

In *The God Delusion* (2006), Dawkins contends that a supernatural creator almost certainly does not exist and that religious faith is a delusion.

Dawkins has been awarded many prestigious academic and writing awards, and he makes regular television, radio, and Internet appearances, predominantly discussing his books, his atheism, and his ideas and opinions as a public intellectual (Creative Commons Legal Code CC0 1.0 Universal).

Deepak Chopra

Chopra studied medicine in India before emigrating to the United States in 1970 where he completed residencies in internal medicine and endocrinology. As a licensed physician, he became chief of staff at the New England Memorial Hospital (NEMH) in 1980. He met Maharishi Mahesh Yogi in 1985 and became involved with the Transcendental Meditation movement (TM). He resigned his position at NEMH shortly thereafter to establish the Maharishi Ayurveda Health Center.

Chopra gained a following in 1993 after he was interviewed on The Oprah Winfrey Show about his books. He then left the TM movement to become the executive director of Sharp HealthCare's Center for Mind-Body Medicine, and in 1996 he co-founded the Chopra Center for Wellbeing. Chopra believes that a person may attain "perfect health," a condition "that is free from disease, that never feels pain," and "that cannot age or die."

Seeing the human body as being undergirded by a "quantum mechanical body" composed not of matter but of energy and information, he believes that "human aging is fluid and changeable; it can speed up, slow down, stop for a time, and even reverse itself," as determined by one's state of mind. He claims that his practices can also treat chronic disease. As of 2014, Deepak Chopra lived in a "health-centric" condominium in Manhattan. (Creative Commons Legal Code CC0 1.0 Universal).

Sam Harris

Harris's first book, *The End of Faith* (2004), won the PEN/Martha Albrand Award for First Nonfiction and remained on *The New York Times* Best Seller list for 33 weeks. In *The Moral Landscape* (2010), he argues that science answers moral problems and can aid human well-being. He then published a long-form essay Lying in 2011, the short book *Free Will* in 2012, *Waking Up: A Guide to Spirituality Without Religion* in 2014, and, with British writer Maajid Nawaz, *Islam and the Future of Tolerance: A Dialogue* in 2015. Harris's work has been translated into over 20 languages.

In September 2013, Harris began releasing the Waking Up podcast, in which he interviews guests, responds to critics, and discusses his views (Creative Commons Legal Code CC0 1.0 Universal).

Sir Roger Penrose

Penrose made contributions to the mathematical physics of general relativity and cosmology. He has received several prizes and awards, including the 1988 Wolf Prize for physics, which he shared with Stephen Hawking for the Penrose–Hawking singularity theorems.

Born in Colchester, Essex, Roger Penrose is a son of psychiatrist and geneticist Lionel Penrose and Margaret Leathes, and the grandson of the physiologist John Beresford Leathes and his Russian wife, Sonia Marie Natanson, who had left St. Petersburg in the late 1880s. His uncle was artist Roland Penrose, whose son with photographer Lee Miller is Antony Penrose. Penrose is the brother of physicist Oliver Penrose and of chess Grandmaster Jonathan Penrose.

Penrose attended University College School and University College, London, where he graduated with a first-class degree in mathematics. In 1955, while still a student, Penrose reintroduced the E. H. Moore generalized matrix inverse, also known as the Moore–Penrose inverse, after it had been reinvented by Arne Bjerhammar in 1951.

Having started research under the professor of geometry and astronomy, Sir W. V. D. Hodge, Penrose finished his Ph.D. at Cambridge in 1958, with a thesis on "tensor methods in algebraic geometry" under algebraist and geometer John A. Todd. He devised and popularized the Penrose triangle in the 1950s, describing it as "impossibility in its purest form," and exchanged material with the artist M. C. Escher, whose earlier depictions of impossible objects partly inspired it. Escher's Waterfall and Ascending and Descending were in turn inspired by Penrose (Creative Commons Legal Code CC0 1.0 Universal).

Stewart Hameroff

Hameroff received his BS degree from the University of Pittsburgh and his MD degree from Hahnemann University Hospital, where he studied before it became part of the Drexel University College of Medicine. He took an internship at the Tucson Medical Center in 1973.

From 1975 onwards, he has spent the whole of his career at the University of Arizona, becoming a professor in the Department of Anesthesiology and Psychology and director for the Center for Consciousness Studies, both in 1999, and finally Emeritus professor for Anesthesiology and Psychology in 2003.

At the very beginning of Hameroff's career, while he was at Hahnemann, cancer-related research work piqued his interest in the part played by microtubules in cell division and led him to speculate that they were controlled by some form of computing. It also suggested to him that part of the solution of the problem of consciousness might lie in understanding the operations of microtubules in brain cells, operations at the molecular and supramolecular level.

These ideas are discussed in Hameroff's first book *Ultimate Computing* (1987). The main substance of this book dealt with the scope for information processing in biological tissue and especially in microtubules and other parts of the cytoskeleton. Hameroff argued that these subneuronal cytoskeleton components could be the basic units of processing rather than the neurons themselves. The book was primarily concerned with information processing, with consciousness secondary at this stage (Creative Commons Legal Code CC0 1.0 Universal).

Rupert Sheldrake

Sheldrake's morphic resonance conjecture posits that "memory is inherent in nature" and that "natural systems, such as termite colonies, or pigeons, or orchid plants, or insulin molecules, inherit a collective memory from all previous things of their kind." Sheldrake proposes that it is also responsible for "telepathy-type interconnections between organisms." His advocacy of the idea encompasses paranormal subjects such as precognition, telepathy and the psychic staring effect as well as unconventional explanations of standard subjects in biology such as development, inheritance, and memory.

Morphic resonance is not accepted by the scientific community as a measurable phenomenon and Sheldrake's proposals relating to it have been characterized as pseudoscience. Critics cite a lack of evidence for morphic resonance and an inconsistency between the idea and data from genetics and embryology. They also express concern that popular attention paid to Sheldrake's books and public appearances undermines the public's understanding of science.

Sheldrake's ideas have found support in the New Age movement from individuals such as Deepak Chopra (Creative Commons Legal Code CC0 1.0 Universal).

Terence McKenna

McKenna formulated a concept about the nature of time-based on fractal patterns he claimed to have discovered in the I Ching, which he called novelty theory, proposing this predicted a transition of consciousness in the year 2012. His promotion of novelty theory and its connection to the Maya calendar is credited as one of the factors leading to the widespread beliefs about 2012 eschatology. Novelty theory is considered pseudoscience (Creative Commons Legal Code CC0 1.0 Universal).

Michael Shermer

Shermer is producer and co-host of the 13-hour Fox Family television series Exploring the Unknown, which was broadcast in 1999. From April 2001 to January 2019, he was a monthly contributor to Scientific American magazine with his Skeptic column. He is also a scientific advisor to the American Council on Science and Health (ACSH).

Shermer was once a fundamentalist Christian but ceased to believe in the existence of God during his graduate studies. He accepts the labels agnostic, nontheist, atheist and others. He has expressed reservations about such labels for his lack of belief in a God, however, as he sees them being used in the service of "pigeonholing," and prefers to simply be called a skeptic. He also describes himself as an advocate for humanist philosophy as well as the science of morality (Creative Commons Legal Code CC0 1.0 Universal).

Leonard Mlodinow

Mlodinow is known to a wider audience through his books for the general public, five of which have been *New York Times* bestsellers, including *The Drunkard's Walk: How Randomness Rules Our Lives*, which was chosen as a *New York Times* notable book, and short-listed for the Royal Society Science Book Prize; *The Grand Design*, co-authored with Stephen Hawking, which argues that invoking God is not necessary to explain the origins of the universe; *War of the Worldviews*, co-authored with Deepak Chopra; and *Subliminal: How Your Unconscious Mind Rules Your Behavior*, which won the 2013 PEN/E.O. Award for Literary Science Writing.

He is also known through his public lectures and media appearances on programs ranging from Morning Joe to Through the Wormhole, and for debating Deepak Chopra on ABC's Nightline (Creative Commons Legal Code CC0 1.0 Universal).

Ralph Abraham

Abraham earned his Ph.D. from the University of Michigan in 1960 and held positions at UC Santa Cruz, Berkeley, Columbia, and Princeton. He has also held visiting positions in Amsterdam, Paris, Warwick, Barcelona, Basel, and Florence.

Abraham founded the Visual Math Institute at UC Santa Cruz in 1975, at that time called the "Visual Mathematics Project." He is editor of World Futures and for the International Journal of Bifurcations and Chaos. He is a member of cultural historian William Irwin Thompson's Lindisfarne Association.

Abraham has been involved in the development of dynamical systems theory since the 1960s and 1970s. He has been a consultant on chaos theory and its applications in numerous fields, such as medical physiology, ecology, mathematical economics, psychotherapy, etc.

Another interest of Abraham concerns alternative ways of expressing mathematics, for example, combining visual and oral formats. He has staged performances in which mathematics, visual arts, and music are combined into one presentation.

Abraham developed an interest in "Hip" activities in Santa Cruz in the 1960s and has a website gathering information on the topic. He credits his use of the psychedelic drug DMT for "swerving his career toward a search for the connections between mathematics and the experience of the Logos"(Creative Commons Legal Code CC0 1.0 Universal).

Joseph (Joe) Rogan

Rogan is an American stand-up comedian, mixed martial arts (MMA) color commentator, podcast host, businessman and former television host and actor.

The Joe Rogan Experience has featured many members of what has been dubbed the Intellectual Dark Web. The study "Alternative Influence: Broadcasting the Reactionary Right on YouTube" identified Rogan as the figure in the alternative influencer network with the highest subscriber count.

Rogan is an avid hunter and is part of the "Eat What You Kill" movement, which attempts to move away from factory farming and the mistreatment of animals raised for food.

He has stated that his personal experiences with meditation in isolation tanks have helped him explore the nature of consciousness as well as improve his performance in various physical and mental activities and overall well-being (Creative Commons Legal Code CC0 1.0 Universal).

References

After 20 years, Stanford encyclopedia of philosophy thrives on the web. (2015, Jul 31). Targeted News Service Retrieved from https://search-proquest-com.contentproxy.phoenix.edu/docview/1700444893?accountid=35812

Agis, d. f. (2014). ser y acción social en el horizonte de la reflexión acerca del tiempo desarrollada por heidegger en ser y tiempo/Being and social action in the horizon of heidegger's thinking on temporality in being and time. Alpha, (39), 219.

Alexander, D. E. (2012). My proof of Heaven: When a neurosurgeon found himself in a coma, he experienced things he never thought possible—a journey to the afterlife. New York: Newsweek Media Group Inc.

Alexander, E. (2013, November 20). The Nature of Consciousness [Video file]. Retrieved from Attribution 2.0 Generic (CC BY 2.0) website: http://www.westminsterforum.org/forum/the-nature-of-consciousness/

Alexander, E., & Newell, K. (2017). Living in a mindful universe. New York, New York: Rodale.

Andersen, R. (2019, March). Scientists are totally rethinking animal cognition. The Atlantic, 1-30. The Atlantic Magazine.

———. (2019, March). Scientists Are Totally Rethinking Animal Cognition What science can tell us about how other creatures experience the world. The Atlantic, (1). Retrieved from https://www.theatlantic.com/magazine/archive/2019/03/what-the-crow-knows/580726/

Art. (2019). In Webster. Retrieved from https://www.merriam-.com/

At Eternity's Gate. (2018, November). YouTube.

Bakalar, N. (2015, June 19). 37.2 Trillion: Galaxies or Human Cells?. New York Times. Retrieved from https://www.nytimes.com/2015/06/23/science/37-2-trillion-galaxies-or-human-cells.html

Barnes, J.& Griffin, F., (1999). Philosophia Togata: Plato and Aristotle at Rome. II. Clarendon Press. ISBN 978-0-19-815222-4.

Behavior research; findings from diagnostica stago provides new data on behavior research (A comparison of procedures for unpairing conditioned reflexive motivating operations). (2018, Apr 21). Psychology & Psychiatry Journal

Berlinski, D. (2009). The devil's delusion, Atheism and its scientific pretensions. New York, NY: Basic Books.

Bobson, D. (2019, March 24). This is what it's like to wake up during surgery. QUARTZ. Retrieved from https://qz.com/1577728/this-is-what-its-like-waking-up-during-surgery/ Attribution-ShareAlike 3.0 Unported (CC BY-SA 3.0)

Bruntrup, G., & Jaskolla, L. (Eds.), Panpsychism: Contemporary Perspectives. : Oxford University Press,. Retrieved 15 Dec. 2018,

Carroll, S. (2016). The big picture. New York, New York: Dutton.

CBS News. (2018). Sunday Morning. Retrieved from https://www.cbsnews.com/news/titanic-the-untold-story/

Choi, C.Q. (2019). Science & Astronomy. Retrieved from https://www.space.com/revealing-universe-before-big-bang.html

Chopra, D. (1994). The seven spiritual laws of success. San Rafael, CA: Amber-Allen.

Chopra, D., & Mlodinow, L. (2012). War of the worldviews, Where science and spirituality meet—and do not. New York, New York: Three Rivers Press—Random House.

Columbia University. (2016). The Big Picture. Retrieved from https://www.math.columbia.edu/~woit/wordpress/?p=8424

Cosmology Today (2015). Michio Kaku vs Richard Dawkins Debate. Retrieved from https://www.youtube.com/watch?v=dmkrI-K7yBo

Cox, B. (2015, February 11). Chopra – CONAN on TBS [Video file]. Retrieved from Conan O'Brien website: http://teamcoco.com/video

Creative Commons Attribution-ShareAlike License. (2018). Deepak Chopra. Retrieved from https://en.wikipedia.org/wiki/Deepak_Chopra

Daily Grail. (2017). Cardiac surgeon tells of veridical near-death experience. Retrieved from https://www.dailygrail.com/2014/02/heart-surgeon-confirms-near-death-experience-account-that-challenges-modern-science/

Dawkins, R. (2006). THE GOD DELUSION. New York, NY: Houghton Miffion Harcourt.

Dawson Ellison, S. G. (2014). Until the resurrection: Poetic fame and the future of the body in english renaissance lyric, 1590-1641 (Order No. 3731502). Available from ProQuest Dissertations & Theses Global. (1734393032).

Donadi, S., Bassi, A., Ferialdi, L., & Curceanu, C. (2013). The effect of spontaneous collapses on neutrino oscillations. Foundations of Physics, 43(9), 1066-1089. doi:10.1007/s10701-013-9732-6

Elmhirst, S. (Year 2015, June 9). Is Richard Dawkins destroying his reputation? THE GUARDIAN. Retrieved from https://www.theguardian.com/science/2015/jun/09/is-richard-dawkins-destroying-his-reputation

Fandom. (December 6, 2018). The Doctor Who Wiki. Retrieved from http://tardis.wikia.com/wiki/Romana_I

Fenker, H. (2019). Jefferson Lab Q&A. Retrieved from https://education.jlab.org/qa/atomicstructure_05.html

Fox, M., & Sheldrake, R. (1996). The physics of Angeles, Exploring the realm where science and spirit meet . San Francisco, CA: Harper Collins.

Gödel's incompleteness theorems (1931). Retrieved from https://en.wikipedia.org/wiki/G%C3%B6del%27s_incompleteness_theorems

Good Reads. (2019). C.S. Lewis. Retrieved from https://www.goodreads.com/quotes/6439-if-we-find-ourselves-with-a-desire-that-nothing-in

Greenblatt, S. (2008). The swerve, How the world became modern. New York, New York: Norton.

Guthrie, K.S., Fideler, D.R., & Joscelyn, G. (1940). The Pythagorean sourcebook and library. Grand Rapids, MI: Phanes Press.

Hameroff, S. (speaker). (2013r, July 15). Consciousness [Audio podcast]. Retrieved from The Chopra Well website: https://www.youtube.com/watch?v=erSd5xep30w

Hameroff, S., & Marcer, P. (1998). Quantum Computation in Brain Microtubules? The Penrose—Hameroff 'Orch OR' Model of Consciousness [and Discussion]. Philosophical Transactions: Mathematical, Physical and Engineering Sciences, 356(1743), 1869–1896. Retrieved from http://www.jstor.org/stable/55017

Harris, S. (Speaker). (2014, September 5). Joe Rogan Experience # 543 [Audio podcast]. Retrieved from Joe Rogan website: http://tu.be/aAnlBW5INYg

Hartdegen, S. J., Peifer, C. J., & Gignac, F. T. (2003). New American Bible. In New Catholic Encyclopedia (2nd ed., Vol. 10, pp. 275–277). Detroit, MI: Gale.

Holt, J. (2018). When Einstein walked with Gödel—Excursions to edge of thought". New York, NY: Macmillan. https://www.youtube.com/watch?v=RYjBXyJu-ME

Juan José Garrido Periñán. (2018). Phenomenological horizons of the spatiality in being and time. Eidos : Revista De Filosofía De La Universidad Del Norte, (29), 150.

Kerr, C. (2015, December 2). I see dead people: Dreams and visions of the dying [Video file]. Retrieved from TED website: http://ted.com/tedx

Klein, J. (2016). How to become a shadow angel in the morning dew.

Koch, C. (2014). Is consciousness universal?. Retrieved from https://www.scientificamerican.com/article/is-consciousness-universal/

Koehne, S. (2018). Religion in the early Nazi milieu: Towards a greater understanding of 'Racist culture'. Journal of Contemporary History, 53(4), 667-691. doi:10.1177/0022009416669420

Lutsyshyna, O. (2012). Classical Samkhya on the authorship of the Vedas. Journal of Indian Philosophy, 40(4), 453-467. doi:10.1007/s10781-012-9161-4

Macisaac, T. (2014). How Common Are Near-Death Experiences? Ndes by the numbers. Retrieved from https://www.theepochtimes.com/how-common-are-near-death-experiences-ndes-by-the-numbers_757401.html

Markides, K. C. (2014). The great partnership: Science, religion, and the search for meaning. Journal of Transpersonal Psychology, 46(1), 126-129.

Martini, R. (2019, April). Do people hear us in Heaven. QUORA, Creative Commons (Open Source), Retrieved from https://www.quora.com/

Marzluff, J.M., & Angel, T. (2005). In the company of crows and ravens. London, UK: Henry Holt.

Mayo Clinic. (2019). General anesthesia. Retrieved from http://https://www.mayoclinic.org/tests-procedures/anesthesia/about/pac-20384568

McNamara, P. (2019). Psychology Today. Retrieved from https://www.psychologytoday.com/us/blog/dream-catcher/201606/can-two-people-have-the-same-dream

Merleau-Ponte, M. (1996). Phenomenology of perception. India: Motilal Banarsidass

Mullin, T. (2012). The cultural practices of modern Chinese Buddhism: Attuning the dharma, by francesca tarocco. Relegere: Studies in Religion and Reception, 2(1), 191-195. doi:10.11157/rsrr2-1-522

Nee, S. (2005). The great chain of being. Nature, 435(7041), 429-429. doi:10.1038/435429a

Parry, D. (2009). Moonshot: The Inside Story of Mankind's Greatest Adventure. New York, NY: Random House.

Penrose, R. (2016). The emperor's new mind. Oxford, UK: OUP Oxford.

———. (speaker). (2018, December 18). Joe Rogan Experience [Audio podcast]. Retrieved from Dec 18, 2018 website: https://www.youtube.com/watch?v=GEw0ePZUMHA

Photon. (2008). In K. L. Lerner & B. W. Lerner (Eds.), The Gale Encyclopedia of Science (4th ed., Vol. 4, pp. 3315-3316). Detroit, MI: Gale. Retrieved from http://link.galegroup.com.content proxy.phoenix.edu/apps/doc/CX2830101780/GVRL?u=uphoenix_uopx&sid=GVRL&xid=3ef6ecb0

Pikovski, I., Vanner, M. R., Aspelmeyer, M., Kim, M. S., & Brukner, C. (2012). Probing Planck-scale physics with quantum optics. Nature Physics, 8(5), 393-397. doi: http://dx.doi.org.contentproxy.phoenix.edu/10.1038/nphys2262

PineCrest. (2019). P-26. Retrieved from https://www.pinterest.com/pin/189643834292346695/?lp=true

ProQuest. (2019). ProQuest Research. Retrieved from https://www.proquest.com/

PURCH. (2019). LiveScience. Retrieved from https://www.livescience.com/65029-dueling-reality-photons.html

Reicher, M. (Winter 2016 Edition), "Nonexistent Objects", The Stanford Encyclopedia of Philosophy Edward N. Zalta (ed.), https://plato.stanford.edu/archives/win2016/entries/nonexistent-objects/.

Reicher, Maria, "Nonexistent Objects", The Stanford Encyclopedia of Philosophy (Winter 2016 Edition), Edward N. Zalta (ed.), https://plato.stanford.edu/archives/win2016/entries/nonexistent-objects/.

Reuter, M., & Saueressig, F. (2012). Quantum Einstein gravity. New Journal of Physics, 14(5), 55022. doi:10.1088/1367-2630/14/5/055022

Rosenblum, B., & Kuttner, F. (2011). Quantum enigma (2nd ed.). New York, New York: Oxford.

Rothman, J. (2014, April). Is Heidegger Contaminated by Nazism? New Yorker.

Ruby, L. (2014, December 6). Nahtoderlebnis im OP [Video file]. Retrieved from Dental Mastermind Group website: http:// https://www.youtube.com/watch?v=8CkEKmCWDC8&t=2s

Rupert Sheldrake (2018). Retrieved from https://www.sheldrake.org/about-rupert-sheldrake.

Sarewitz, D. (2015). Gaia. In J. B. Holbrook (Ed.), Ethics, Science, Technology, and Engineering: A Global Resource (2nd ed., Vol. 2, pp. 317-318). Farmington Hills, MI: Macmillan Reference USA.

Science. (2019). In Hard and Soft Science. Retrieved from https://en.wikipedia.org/wiki/Hard_and_soft_science.

Selk, A. (2019, February 5). Harvard's top astronomer says an alien ship may be among us — and he doesn't care what his colleagues think. The Washington Post. Retrieved from https://www.msn.com/en-us/news/us/harvards-top-astronomer-says-an-alien-ship-may-be-among-us-—-and-he-doesnt-care-what-his-colleagues-think/ar-BBTaWxw?ocid=spartandhp.

Sheldrake, R. (2009). Morphic resonance, The nature of formative causation (3rd ed.). Toronto, CA: Park Street Press.

Sheldrake, R., & Shermer, M. (2016). Arguing science. Rhinebeck, NY: Monkfish.

Shermer, M. (2012, October 22). Michael Shermer fires back at Deepak Chopra | WHO ARE YOU? Pt 3 – Deepak Chopra [Video file]. Retrieved from YouTube website: https://www.youtube.com/watch?v=NRqxnI96aIE

Shermer, M. (2015). The moral arc. New York, NY: St Martin's Griffin.

Singer, I.B. (1980) The séance and other stories. New York: Farrar, Straus & Giroux, p. 217

Skeptic (2019). Retrieved from https://g.co/kgs/Z1paoq

Sones, B., & Sones, R. (2015). Strange but true: 'speciesist' holds humans above all. The Berkshire Eagle.

Spanton, T. (2009, Jul 18). One giant step for mankind, one small leak for man. The Sun [London (UK)] 18 July 2009: 22.

Stanford University Library. (2019). Spinoza. Retrieved from https://plato.stanford.edu/entries/spinoza/

The guardian (2019). Retrieved from https://www.theguardian.com/science/2009/apr/02/paul-dirac-strangest-man-farmelo-quantum.

Tyson, N. (2017). Astrophysics for people on the move. New York, New York: Norton.

Volk, S. (n.d.). . Down the quantum rabbit hole, (), . Retrieved from http://discovermagazine.com/bonus/quantum.

Volk, S. (nd, nd). Down the quantum rabbit hole. Discovery, nd(nd), 1-97. Retrieved from http://discovermagazine.com/bonus/quantum.

Water. (2019). In (www.cyber-nook.com/water/p-quotes.htm). Retrieved from http://www.cyber-nook.com/water/p-quotes.htm.

Weisberg, J. (2019). The Internet Encyclopedia of Philosophy. Retrieved from https://www.iep.utm.edu/hard-con/

Wheeler, Michael, "Martin Heidegger", The Stanford Encyclopedia of Philosophy (Winter 2018 Edition), Edward N. Zalta (ed.), https://plato.stanford.edu/archives/win2018/entries/heidegger/.

Wikipedia. (2019). Doctor Who. Retrieved from https://en.wikipedia.org/wiki/Doctor_Who.

———. (2019). Lucretius. Retrieved from https://en.wikipedia.org/wiki/Lucretius.

———. (2019). Roger Penrose. Retrieved from https://en.wikipedia.org/wiki/Roger_Penrose.

———. (2019). Sam Harris. Retrieved from https://en.wikipedia.org/wiki/Sam_Harris.

———. (2019). Yang. Retrieved from https://en.wikipedia.org/wiki/Yang_(surname).

Wolpe, D. (2009). Why faith matters. New York, New York: Harper Collins.

YouTube (2015). Retrieved from https://www.youtube.com/watch?v=dmkrI-K7yBo.

Zhao, c. (2014). Conflicts between ambition and love of ralph in the thorn birds in light of maslow's hierarchy of needs. Journal of Arts and Humanities, 3(8), 104.

Explanations

Quantum entanglement is a physical phenomenon that occurs when pairs or groups of particles are generated, interact, or share spatial proximity in ways such that the quantum state of each particle cannot be described independent of the state of the others, even when the particles are separated by a large distance.

Measurements of physical properties such as position, momentum, spin, and polarization, performed on entangled particles, are found to be correlated.

For example, if a pair of particles is generated in such a way that their total spin is known to be zero, and one particle is found to have clockwise spin on a certain axis, the spin of the other particle, measured on the same axis, will be found to be counterclockwise, as is to be expected due to their entanglement.

However, this behavior gives rise to seemingly paradoxical effects: any measurement of a property of a particle performs an irreversible collapse on that particle and will change the original quantum state. In the case of entangled particles, such a measurement will be on the entangled system as a whole.

Such phenomena were the subject of a 1935 paper by Albert Einstein, Boris Podolsky, and Nathan Rosen, and several papers by Erwin Schrödinger shortly thereafter, describing what came to be known as the EPR paradox. Einstein and others considered such behavior to be impossible, as it violated the local realism view of causality (Einstein referring to it as "spooky action at a distance") and argued that the accepted formulation of quantum mechanics must, therefore, be incomplete.

Later, however, the counterintuitive predictions of quantum mechanics were verified experimentally in tests where the polarization or spin of entangled particles were measured at separate locations, statistically violating Bell's inequality. In earlier tests it couldn't be absolutely ruled out that the test result at one point could have been subtly transmitted to the remote point, affecting the outcome at the second location. However, so-called "loophole-free" Bell tests have been performed in which the locations were separated such that communications at the speed of light would have taken longer—in one case 10,000 times longer—than the interval between the measurements.

According to some interpretations of quantum mechanics, the effect of one measurement occurs instantly. Other interpretations which don't recognize wave function collapse dispute that there is any "effect" at all. However, all interpretations agree that entanglement produces correlation between the measurements and that the mutual information between the entangled particles can be exploited, but that any transmission of information at faster-than-light speeds is impossible.

Quantum entanglement has been demonstrated experimentally with photons, neutrinos, electrons, molecules as large as Buckyballs, and even small diamonds. The utilization of entanglement in communication and computation is a very active area of research (cc Attribution-ShareAlike 3.0 Unported (CC BY-SA 3.0)).

The Planck Constant (denoted h, also called Planck's constant) is a physical constant that is the quantum of electromagnetic action, which relates the energy carried by a photon to its frequency. A photon's energy is equal to its frequency multiplied by the Planck constant. The Planck constant is of fundamental importance in quantum mechanics, and in metrology, it is the basis for the definition of the kilogram.

At the end of the 19th century, physicists were unable to explain why the observed spectrum of black body radiation, which by then had been accurately measured, diverged significantly at higher frequencies from that predicted by existing theories.

In 1900, Max Planck empirically derived a formula for the observed spectrum. He assumed that a hypothetical electrically charged oscillator in a cavity that contained black body radiation could only change its energy in a minimal increment, E, that was proportional to the frequency of its associated electromagnetic wave. He was able to calculate the proportionality constant, h, from the experimental measurements, and that constant is named in his honor.

In 1905, the value E was associated by Albert Einstein with a "quantum" or minimal element of the energy of the electromagnetic wave itself. The light quantum behaved in some respects as an electrically neutral particle, as opposed to an electromagnetic wave. It was eventually called a photon.

Max Planck received the 1918 Nobel Prize in Physics "in recognition of the services he rendered to the advancement of Physics by his discovery of energy quant."

Since energy and mass are equivalent, the Planck constant also relates mass to frequency. By 2017, the Planck constant had been measured with sufficient accuracy in terms of the SI base units, that it was central to replacing the metal cylinder, called the International Prototype of the Kilogram (IPK) that had defined the kilogram since 1889. The new definition was unanimously approved at the General Conference on Weights and Measures (CGPM) on 16 November 2018 as part of the 2019 redefinition of SI base units.

For this new definition of the kilogram, the Planck constant, as defined by the ISO standard, was set to 6.62607015×10−34 J·s exactly. The kilogram was the last SI base unit to be re-defined by a fundamental physical property to replace a physical artifact. (Creative Commons 2019) https://en.m.wikipedia.org/wiki/Planck_constant) (Creative Commons Legal Code CC0 1.0 Universal). (cc Attribution-ShareAlike 3.0 Unported (CC BY-SA 3.0)).

Quantum Decoherence is the loss of quantum coherence. In quantum mechanics, particles such as electrons are described by a wave function, a mathematical representation of the quantum state of a system; a probabilistic interpretation of the wave function is used to explain various quantum effects. As long as there exists a definite phase relation between different states, the system is said to be coherent.

Coherence is preserved under the laws of quantum physics, and this is necessary for the functioning of quantum computers. If a quantum system is perfectly isolated, it would be impossible to manipulate or investigate it. If it is not perfectly isolated, for example, during a measurement, coherence is shared with the environment and appears to be lost with time, a process called quantum decoherence. As a result of this process, quantum behavior is apparently lost, just as energy appears to be lost by friction in classical mechanics.

Decoherence was first introduced in 1970 by the German physicist H. Dieter Zeh and has been a subject of active research since the 1980s. Decoherence has been developed into a complete framework, but it does not solve the measurement problem, as the founders of decoherence theory admit in their seminal papers.

Decoherence can be viewed as the loss of information from a system into the environment (often modeled as a heat bath), since every system is loosely coupled with the energetic state of its surroundings. Viewed in isolation, the system's dynamics are non-unitary (although the combined system plus environment evolves in a unitary fashion). Thus, the dynamics of the system alone are irreversible. As with any coupling, entanglements are generated between the system and environment. These have the effect of sharing quantum information with—or transferring it to—the surroundings.

Decoherence has been used to understand the collapse of the wave function in quantum mechanics. Decoherence does not generate actual wave-function collapse. It only provides an explanation for the observation of wave-function collapse, as the quantum nature of the system "leaks" into the environment. That is, components of the wave function are decoupled from a coherent system and acquire phases from their immediate surroundings. A total superposition of the global or universal wave function still exists (and remains coherent at the global level), but its ultimate fate remains an interpretational issue.

Specifically, decoherence does not attempt to explain the measurement problem. Rather, decoherence provides an explanation for the transition of the system to a mixture of states that seem to correspond to those states observers perceive. Moreover, our observation tells us that this mixture looks like a proper quantum ensemble in a measurement situation, as we observe that measurements lead to the "realization" of precisely one state in the "ensemble."

Decoherence represents a challenge for the practical realization of quantum computers, since such machines are expected to rely heavily on the undisturbed evolution of quantum coherences. Simply put, they require that coherent states be preserved and that decoherence is managed, in order to actually perform quantum computation (cc Attribution-ShareAlike 3.0 Unported [CC BY-SA 3.0]).

Robson (2019), Donna Penner's Traumatic Memories

The following article first appeared on Mosaic and is republished here under a Creative Commons license.

Robson, D. (2019). MOSAIC. Creative Commons Attribution 4.0 license (CC BY 4.0). Retrieved fromhttps://mosaicscience.com/story/anaesthesia-anesthesia-awake-awareness-surgery-operation-or-paralysed/republish/(Creative Commons Legal Code CC0 1.0 Universal).

It can be the smallest event that triggers Donna Penner's traumatic memories of an operation from more than ten years ago.

One day, for instance, she was waiting in the car as her daughter ran an errand and realized that she was trapped inside. What might once have been a frustrating inconvenience sent her into a panic attack. "I started screaming. I was flailing my arms, I was crying," she says. "It just left me so shaken."

Even the wrong clothing can make her anxiety worse. "Anything that's tight around my neck is out of the question because it makes me feel like I'm suffocating," says Donna, a 55-year-old from Altona in Manitoba, Canada.

Donna would not be like this if it were not for a small medical procedure that she had before her 45th birthday. She was working in the accountancy department of a local trucking company and had just celebrated the wedding of one of her daughters. But she had been having severe bleeding and pain during her period, and her family physician had suggested that they investigate the causes with exploratory surgery.

It should have been a routine procedure, but, for reasons that are far from clear, the general anesthetic failed. Rather than lying in peaceful oblivion, she woke up just before the surgeon made the first cut into her abdomen. With her body still paralyzed by the anesthetic drugs, she was unable to signal that anything was wrong.

So she remained frozen and helpless on the operating table as the surgeon probed her body, while she experienced indescribable agony. "I thought, 'This is it, this is how I'm going to die, right here on the table, and my family will never know what my last few hours were like because no one's even noticing what's going on.'"

The lingering trauma can resurface with the slightest trigger, and still causes her to have "two or three nightmares each night." Having been put on medical leave from her job, she has lost her independence. She suspects that she will never fully escape the effects of that day more than a decade ago. "It's a life sentence."

For years, anesthesia awareness has been shrouded in mystery. Although extreme experiences like Donna's are rare, there is now evidence that around five percent of people may wake up on the operating table—and possibly many more.

Thanks to the amnesiac effects of the drugs, however, most of these people will be unable to remember anything about the event—and whether or not that is something we should be concerned about is both a practical and a philosophical question.

These results are all the more significant given just how often general anesthesia is now used. "Almost three million general anesthetics happen each year in the UK alone," says Peter Odor, a registrar at St George's Hospital in London. "As a consequence, it is more probable than not that someone, somewhere in the world, right now is aware during their surgery."

We once knew surprisingly little about why anesthesia works. Now, however, researchers are striving to understand more about the nature of going under and the circumstances in which anesthesia doesn't work, in the hope of making advances that might reduce the risk of anesthesia awareness. And, with a greater understanding of the anesthetized state, we may even be able to turn a rudimentary awareness to our advantage—in the form of medical hypnosis.

Let's be clear: anesthesia is a medical miracle. Since at least the time of Hippocrates, physicians and medicine men had hunted for a good way to ease the pain of medical procedures. While drugs such as alcohol, opium, and even hemlock could act as sedatives, their efficacy was unreliable; most patients did not escape the torture.

By the 1840s, scientists had discovered various gases that appeared to have sedative effects. One of these, sulphuric ether, had attracted the particular attention of a dentist based in Boston, named William Morton, who put it to the test in a public demonstration at the Massachusetts General Hospital in 1846. Although the patient was still able to mutter half-coherent thoughts, he reportedly felt no pain, just the faint sense of his skin being "scratched with a hoe".

The news of the demonstration soon spread throughout the medical establishment, heralding the start of the anesthetic era. With the subsequent discovery of even more effective anesthetic agents such as chloroform, the agony of the surgical knife seemed to be a thing of the past.

Today, anesthetists have a wide range of pain-killing and consciousness-reducing drugs at their disposal, and the exact choice will depend on the procedure and the patient's particular needs.

Often, the aim is not to produce a loss of consciousness but simply to remove the sensation from a particular part of the body. So-called regional anesthetics include spinal and epidural anesthetics, both of which are delivered between the bones of your back to numb the lower half of your body. These are commonly used during childbirth, bladder operations, and hip replacements.

You may also be given a sedative—which produces a relaxed, sleepy state but does not fully eliminate your awareness. General anesthesia, in contrast, aims to do just that, creating an unresponsive drug-induced coma or controlled unconsciousness that is deeper and more detached from reality even than sleep, with no memories of any events during that period. As Robert Sanders, an anesthetist at the University of Wisconsin–Madison, puts it: "We've apparently ablated this period of time from that person's experience."

We still don't know exactly why anesthetic agents dim our consciousness, but they are thought to interfere with various brain chemicals called neurotransmitters. These chemicals turn up or turn down the activity of neurons, particularly the widespread communication between different brain regions.

Propofol, for instance—a milky-white fluid used in general anesthetics and some types of sedation—seems to amplify the effects of GABA, an inhibitor that damps down activity in certain areas of the brain, as well as communication between them. These areas include the frontal and parietal regions, at the front and towards the back of the head.

Sanders's colleagues recently used a form of non-invasive brain stimulation to demonstrate this principle in action, with propofol silencing the waves of activity you would normally see spreading across the brain in response to the stimulation.

"It's very possible that anesthesia interferes with that ascending transmission of information," he says. And without it, the mind temporarily disintegrates, becoming a blank screen with no ability to process or respond to the body's signals.

In the clinic, there are many complicating factors to consider, of course. An anesthetist may choose to use one drug to induce the temporary coma and another to maintain it, and they need to consider many factors—such as the patient's age and weight, whether they smoke or take drugs, the nature of their illness—to determine the doses.

Many procedures also use muscle relaxants. For example, nearly half of general anesthetics administered in the UK included neuromuscular blockers. These drugs temporarily paralyses the body, preventing spasms and reflexes that could interfere with the surgery, without raising the dose of the anesthetic drugs to dangerously high levels.

Neuromuscular blockers can also ease the insertion of a tube through the windpipe, which can be used to ensure the airway remains open as well as to deliver oxygen and drugs, and to prevent stomach acid from entering the lungs. If the paralytic agents also stop muscles in the diaphragm and abdomen from moving, however, the patient's breathing must be assisted artificially with a ventilator.

This all makes anesthesia as much art as science, and in the vast majority of cases, it works astonishingly well. More than 170 years after Morton's public demonstration, anesthetists across the world plunge millions of people each year into comas and then bring them out safely. This doesn't just reduce patients' immediate suffering; many of the most invasive life-saving procedures would simply not be possible without good general anesthesia.

But as with any medical procedure, there can be complicating factors. Some people may have a naturally higher threshold for anesthesia, meaning that the drugs don't reduce the brain's activity enough to dim the light of consciousness.

In some cases, such as injuries involving heavy bleeding, an anesthetist may be forced to use a lower dose of the anesthetic for the patient's own safety.

It may also be difficult to time the effects of the different drugs, to ensure that the so-called induction dose—which gets you to sleep—doesn't fade before the maintenance dose—to keep you unconscious—kicks in.

In some situations, you might be able to raise or lower your limb, or even speak, to show the anesthetic is not working before the surgeon picks up their scalpel. But if you have also been given neuromuscular blockers, that won't be possible. The unfortunate result is that a small proportion of people may lie awake for part or all of their surgery without any ability to signal their distress.

Donna tells me about her own experience, during a lengthy telephone conversation from her home in Canada.

She says that she had felt anxious in the run-up to the operation, but she had had general anesthetic before without any serious problems. She was wheeled into the operating theatre, placed on the operating table, and received the first dose of anesthesia. She soon drifted off to sleep, thinking, "Here I go."

When she woke up, she could hear the nurses buzzing around the table, and she felt someone scrubbing at her abdomen—but she assumed that the operation was over and they were just clearing up. "I was thinking, 'Oh boy, you were anxious for no reason.'" It was only once she heard the surgeon asking the nurse for a scalpel that the truth suddenly dawned on her: the operation wasn't over. It hadn't even begun.

The next thing she knew, she felt the blade of his knife against her belly as he made his first incision, leading to excruciating pain. She tried to sit up and to speak—but thanks to a neuromuscular blocker, her body was paralyzed. "I felt so . . . so powerless. There was just nothing I could do. I couldn't move, couldn't scream, couldn't open my eyes," she says. "I tried to cry just to get tears rolling down my cheeks, thinking that they would notice that and notice that something was going on. But I couldn't make tears."

The frustration was immense. "It felt like someone was sitting on me and holding me down and there was absolutely nothing I could do."

Eventually, she tried to focus all her attention on moving one foot, which she managed to wiggle very slightly—and felt astonishing relief when one of the nurses placed his hand over it. Before she could move it again, however, the nurse had let go. She tried a total of three times, all with the same result. "It was very frustrating for me knowing that was the only way to communicate and it wasn't working."

Donna's torment should have finished after the surgeon had ended his work. But as the neuromuscular blockers began to wear off, she started to move her tongue around the tube stuck down her throat; it was a way, she thought, of signaling to the staff that she was awake.

Unfortunately, the staff misread her attempts at communication, and began to withdraw the tube prematurely, before the paralytic agent had faded enough for her lungs to be able to operate on their own. "So here I was lying on the table and he took away my life support, my oxygen; I could not take a breath," Donna says. She assumed she would die.

At this point, the operating room began to feel more distant, as she felt her mind escape in an out-of-body experience. A committed Christian, she says she felt the presence of God with her. It was only after the staff restored her oxygen supply that she drifted back into the operating room, to wake, crying.

That pain, the fear, the sense of absolute helplessness all still linger to this day—feelings of trauma that have led her to be put on medical leave from her job. And that has meant a loss of independence and confidence, and the abandonment of many of the hopes and dreams that she had built with her husband.

"It's hard to sit at home here and watch all the neighbors hurrying out of their house in the morning, jump in their cars, and go off to work, and I can't."

Various projects around the world have attempted to document experiences like Donna's, but the Anesthesia Awareness Registry at the University of Washington, Seattle, offers some of the most detailed analyses. Founded in 2007, it has now collected more than 340 reports—most from North America—and although these reports are confidential, some details have been published, and they make illuminating reading.

Nearly all the patients included said they heard voices or other sounds under general anesthesia (patients' eyes are typically closed during surgery so visual experiences tend to be less common).

"I heard the type of music and tried to figure why my surgeon chose that," one patient told the registry. "I heard several voices around me.

They seemed to be panicking. I heard them say they were losing me," another reported.

As you might expect, a large majority of the accounts—more than seventy percent—also contain reports of pain. "I felt the sting and burning sensation of four incisions being made, like a sharp knife cutting a finger," wrote one. "Then searing, unbearable pain."

"There were two parts I remember quite clearly," wrote a patient who had had a wide hole made in his femur. "I heard the drill, felt the pain, and felt the vibration all the way up to my hip. The next part was the movement of my leg and the pounding of the 'nail'." The pain, he said, was "unlike anything I thought possible."

It is the paralyzing effects of the muscle blockers that many find most distressing, however. For one thing, it produces the sensations that you are not breathing—which one patient described as "too horrible to endure."

Then there's the helplessness. Another patient noted: "I was screaming in my head things like 'don't they know I'm awake, open your eyes to signal them'."

To make matters worse, all of this panic can be compounded by a lack of understanding of why they are awake but unable to move. "They have no reference point to say, why is this happening?" says Christopher Kent at the University of Washington, who co-authored the paper about these accounts. The result, he says, is that many patients come to fear that they are dying. "Those are the worst of the anesthesia experiences."

Estimates of how often anesthesia awareness happens have varied depending on the methods used, but those relying on patient reports had tended to suggest it was very rare indeed.

One of the largest and most thorough investigations was the fifth National Audit Project carried out by British and Irish anesthetists' associations, in which every public hospital in the UK and Ireland had to report any incidents of awareness for a year. The results, published in 2014, found that the overall prevalence was just 1 in 19,000 patients undergoing anesthesia. The figure was higher—around 1 in 8,000—if the anesthesia included paralyzing drugs, which is to be expected, since they prevent the patient from alerting the anesthetist that there is a problem before it is too late.

These low numbers were comforting news. As the media reported at the time, you were more likely to die during surgery than to awake during it, confirming many doctors' suspicions that this was a very remote risk.

Unfortunately, these figures are probably underestimating, as Peter Odor explains to me at St George's Hospital in London. For one thing, the National Audit Project relied on patients themselves reporting directly to the hospital—but many people may feel unable or unwilling to come forward and would instead prefer to just put the experience behind them.

There are also the amnesiac effects of the drugs themselves. "Anesthetic drugs disrupt your ability to encode a memory," said Odor. "And the dose that you give to obliterate memories is lower than that you need to obliterate consciousness. So memory goes well before consciousness goes."

The result is that many more people might be conscious during surgery, but they simply can't remember it afterwards.

To investigate this phenomenon, researchers are using what they call the isolated forearm technique. During the induction of the anesthesia, the staff place a cuff around the patient's upper arm that delays the passage of the neuromuscular agent through the arm. This means that, for a brief period, the patient is still able to move their hand. So a member of staff could ask them to squeeze their hand in response to two questions: whether they were still aware, and, if so, whether they felt any pain.

In the largest study of this kind to date, Robert Sanders at the University of Wisconsin–Madison recently collaborated with colleagues at six hospitals in the US, Europe and New Zealand. Of the 260 patients studied, 4.6% responded to the experimenters' first question, about awareness. That is hundreds of times greater than the rate of remembered awareness events that had been noted in the National Audit Project. And around four in ten of those patients who did respond with the hand squeeze—1.9% across the whole group—also reported feeling pain in the experimenters' second question.

These results raise some ethical quandaries. "Whenever I talk to the trainees, I talk about the philosophical element to this," says Sanders. "If the patient doesn't remember, is it concerning?"

Sanders says that there's no evidence that the patients who respond during the isolated forearm experiments, but fail to remember the experience later, do go on to develop PTSD or other psychological issues like

Donna. And without those long-term consequences, you might conclude that the momentary awareness is unfortunate, but unalarming.

Yet the study does make him uneasy, and so he conducted a survey to gather the public's views on the matter. Opinions were mixed. "Most people didn't think that amnesia alone is sufficient—but a surprisingly large minority thought that as long as you didn't remember the event, it's OK," Sanders says.

"My view is that the patient is expecting to be unconscious, and, as a researcher who wants to understand the mechanisms at play, but also a clinician who wants to deliver high-quality care and meet the expectations of the patient, we are duty-bound to understand this balance and to find out the true rates and the true impact of those events, whether they have any impact or not, and the ways we can curtail them."

Given that the vast majority of patients will emerge from general anesthesia without traumatic memories, there is the danger that reports of anesthesia awareness—including this one—will needlessly increase anxiety before operations.

In the worst-case scenario, those fears could even prevent someone from having an essential medical procedure. Certainly, anesthetists such as Sanders have emphasized that the risks of explicit recall are small, but if you are anxious you should talk to the hospital staff about your concerns.

But there are, nevertheless, strong arguments for making this phenomenon more widely known. For instance, as the reports from the University of Washington's registry show, some patients' distress was amplified by their lack of understanding of what was happening. They assumed that their awareness was a sign that they were dying. Perhaps if they had known the risk beforehand, that panic might have been assuaged.

A better understanding of anesthesia awareness might also help medical staff to deal with patients who have experienced this trauma. Many—including Donna—have felt that their accounts were misunderstood or simply dismissed by medical professionals.

The Washington registry, for instance, found that 75 percent of those who had reported awareness were unsatisfied with the response, with 51 percent saying that neither the anesthetist nor the surgeon expressed sympathy for their experience. Overall, just 10 percent received an apology, and only 15 percent were referred for counseling to help them to deal with the trauma.

Donna says that many of the staff at her hospital seemed completely bewildered by her trauma. As she came around, she tried to explain to the nurses what had happened to her, but they just stood in silence, she says. "I'll never forget their expressions—it was like they were in shock." She puts this down to a lack of education and understanding of the phenomenon. "They don't know how to handle this kind of a situation."

Having gained strength in the years following the trauma, Donna is now trying to remedy the problem, by working with Canadian universities to educate doctors about the risks of anesthesia awareness and the best ways to treat patients. "I want them to be prepared, because when things go wrong you need to know how you are going to react to the patient, because that is crucial to the patient's recovery process."

The ultimate goal, though, is to prevent these traumatic experiences from occurring in the first place, with studies using the isolated forearm technique helping to identify the best procedures to ensure unconsciousness. "There could be specific combinations of drugs that could produce the right blend of anesthesia to insulate people from the external sensory world a bit better," Sanders says.

There is even a chance that, as our understanding of the anesthetized state deepens, we might be able to turn rudimentary unconscious responsiveness during surgery to our advantage. Certain forms of medical hypnosis have been found to have a real effect on patients' experiences in controlled clinical trials—and the anesthetic state could be the perfect time to put it into practice.

Although widespread signaling across the brain appears to be impaired when people are under general anesthesia, there is evidence that certain areas—including the auditory cortex—remain responsive, suggesting that medical staff might be able to send suggestions and encouragement, while a patient is unconscious, to reduce their pain after surgery.

Studies investigating this possibility are few, but Jenny Rosendahl, at Jena University Hospital in Germany, and colleagues have attempted to gather all the evidence to date. Their meta-analysis showed a small but significant improvement in the patients' ratings of post-operative nausea and vomiting, and less use of morphine after the operation.

Clearly, no one is suggesting that you would keep a patient fully aware on purpose, but perhaps one day more anesthetists will be able to make use of the brain's ability to absorb information on the operating table. It is an exciting thought that the words we hear during this mysterious twilight zone could have a lasting effect on our recovery.

About the Author

John LaCasse's life has been filled with almost every conceivable emotion as his adventurous spirit, ongoing curiosity, devil-may-care attitude, and endless love of nature, music, art, animals, friends and family are open, evident and expressed at a moment's notice. Gratitude for loving parents and a grandmother who introduced him to all of the above; pride in business success; anger witnessing betrayal; both joy and grief in fatherhood; love for devoted partners. If "variety is the spice of life," Dr. John LaCasse has consumed multi-milligrams and attracted followers from every walk of life, every age, every gender.

Along the way there have been times to slow down, look around, peek *inside* to see what should be changed and as a result education made a U-turn in middle age and Transcendental learning and teaching, an MBA in International Business, and a Ph.D. at North Central University took precedence. A Forbes School of Business Faculty Member of the Year award echoed the dozens of kudos from his students.

Memberships in the American Association of University Professors, academic honorariums in Kappa Delta Pi and Golden Key, and an examiner for the International Baccalaureate continue his involvement in education, but creative writing has become a passion with now five books in print, including *Fight for the Quantum*, *After Your Children Die*, and *Deals, Danger, Destiny*, the forerunner of *Floppy Feathers*.

John lives in Seattle, Washington, with his life partner, Christine Burgoyne.

www.ingramcontent.com/pod-product-compliance
Lightning Source LLC
LaVergne TN
LVHW040223110826
845146LV00004B/1263

9798888193747